A.O. Cifuentes

Using MSC/NASTRAN

Statics and Dynamics

With 94 Illustrations

Springer-Verlag
New York Berlin Heidelberg
London Paris Tokyo Hong Kong

Arturo O. Cifuentes
The MacNeal-Schwendler Corporation
Los Angeles, CA, USA

Library of Congress Cataloging-in-Publication Data

Cifuentes, Arturo O.
    Using MSC/NASTRAN : statics and dynamics / Arturo O. Cifuentes.
        p.   cm.
    Includes index.
    1. Structural analysis (Engineering)—Matrix methods—Data
processing. 2. Finite element method. 3. NASTRAN (Computer
program) I. Title.
TA647.C54   1989
624.1'71'01515353—dc20   89-11330

Printed on acid-free paper.

Camera-ready copy prepared by the author using Microsoft Word.

9 8 7 6 5 4 3 2 1

ISBN-13: 978-0-387-97032-5   e-ISBN-13: 978-1-4613-8917-0
DOI: 10.1007/ 978-1-4613-8917-0

This book is dedicated to my best friend and wife Ventura Charlin

UNO

Uno busca lleno de esperanza
el camino que los sueños
prometieron a sus ansias.
Sabe que la lucha es cruel
y es mucha; pero lucha y se desangra
por la fe que lo empecina.

Enrique S. Discépolo y Marianito Mores

Preface

The idea of writing this book came up one night while having dinner with Ventura at the Crocodile Cafe in Pasadena. This was really a joint project, that could have turned into a nightmare without her support, encouragement, and expertise in personal computers. For all these things, and for tolerating my sometimes single-minded attention, I am very grateful to her.

I am also very much indebted to six good friends, Paul Burridge, Mladen Chargin, Gary Dilley, Carl Hennrich, Hector Jensen and Mark Miller, who read the entire manuscript of this book and made many useful suggestions.

I also want to thank Burt Alperson for his guidance and advice during the preparation of this book.

Finally, I thank the Department of Civil Engineering of the University of Southern California for the support provided during the course of this project, and my students of all these years for asking tough questions.

Contents

PART 2
Dynamics

Introduction

At the present time the finite element method is the most important analysis tool in engineering, and MSC/NASTRAN is the most powerful --and most widely used-- finite element program in the world. MSC/NASTRAN is a general purpose code developed and maintained by The MacNeal-Schwendler Corporation, an international organization based in Los Angeles and with offices throughout the world.

MSC/NASTRAN has been used for more than fifteen years by engineers and researchers to solve a wide range of problems in civil, mechanical, and aerospace engineering. The code is mainly a structural analysis tool for statics and dynamics, including linear and nonlinear behavior. It can also be used for heat transfer, aeroelasticity, and a certain class of fluid/structure interaction and electromagnetic problems. This book deals with linear static and dynamic analysis.

MSC/NASTRAN is written in Fortran and has more than 500,000 lines of source code. Its documentation consists of about fifteen thick volumes and is rather threatening for the first-time user who often faces the problem of not knowing where to start. However, in spite of this massive amount of documentation, the code is friendly and easy to use.

The purpose of this book is to help new MSC/NASTRAN users overcome the anxiety and confusion that arises when faced with such a huge amount of documentation, and to bring them to a reasonable level of proficiency in a few weeks. Thus, the book is aimed at the person who has some structural analysis background but knows nothing about MSC/NASTRAN. It must be emphasized that this book does not intend to be a text on finite element theory or structural analysis.

This book grew out of a set of notes developed for a class in computerized structural analysis that the author has been teaching at the University of Southern California for the last three years. The students who take this class are typically seniors or first-year graduate students majoring in civil, mechanical, or aerospace engineering. Most of them have taken a one or two semester class in classical structural analysis and an introductory course in finite element theory.

The text is based on twenty-eight carefully chosen examples that show the most important and fundamental features of MSC/NASTRAN. The reader is advised to go through these examples in sequential order since key points are introduced in each of them. Each example, however, is a self-contained unit illustrating a different type of analysis. This makes the text useful for future reference. It is important to point out that the examples in the book have been purposely selected to be very simple. In fact, many of these problems can be solved without using a computer. The idea is to demonstrate the most important features of MSC/NASTRAN without distracting the reader with complex geometric or loading configurations. However, it is expected that the reader will be able to bridge the gap between these examples and real life applications involving models with several thousand degrees of freedom without too much difficulty. After all, the power of a general program such as MSC/NASTRAN is best appreciated in the analysis of highly complex structures with irregular configurations and subjected to a wide variety of static and dynamic excitations.

After finishing the book, the reader should be able to solve a large number of structural analysis problems that commonly appear in engineering practice. In addition, he will have a general overview of the capabilities offered by MSC/NASTRAN and a practical understanding of the advantages and limitations of the finite element method.

Basic MSC/NASTRAN concepts

In order to run MSC/NASTRAN the user needs to create a file to define the structure under consideration, the loads acting on it, the boundary conditions, the type of analysis to be performed, and the plots to be created. This file is called the input file.

Once this file has been prepared, the user is ready to execute MSC/NASTRAN by invoking a special computer command followed by the name of the input file. This command will typically be NASTRAN, NAST65, NAST66, or something similar. Consult the person in charge of your computer to find out the appropriate keyword.

After typing this keyword (say, NASTRAN) followed by the name of the input file the computer will execute MSC/NASTRAN. Upon completion of the finite element calculations, MSC/NASTRAN creates several files each containing different types of information. The file that contains the results (in the general sense of the word) is called the output file. Another important file is the plot file. This file contains the information needed to create plots that show different aspects of the problem under study. It is normally required to use a graphics terminal to create these plots.[1]

---

[1] Some readers will probably prefer to use a post-processor to plot their results. If this is the case, simply ignore the comments explaining how to obtain MSC/NASTRAN plots.

A few other things that need to be clear:

1   MSC/NASTRAN does not have a built-in system of units. Therefore, it is up to the user to make sure that the units used to input the data are consistent. If the user mixes units -- for example the cross section of a beam in square meters and the mass density of a material in units of mass per cubic foot -- MSC/NASTRAN will produce meaningless numbers without any warnings.

2   There are two basic entities in a finite element model: the finite elements themselves and the nodal points to which the elements are connected. The finite elements in MSC/NASTRAN have different names depending on their features. A four-node quadrilateral plate element, for example, is called QUAD4. A nodal point (or node) of the finite element model is called GRID or GRID point.

3   Each GRID point has (in principle) six degrees of freedom: three translations and three rotations. The three translations are denoted as 1, 2, and 3. Hence, 1 corresponds to the translation (or displacement) in the x-direction; 2 corresponds to the translation in the y-direction; and 3 corresponds to the translation in the z-direction.[2] Rotations about x, y, and z are denoted as 4, 5, and 6, respectively.

4   MSC/NASTRAN has a built-in coordinate system called the basic coordinate system. This is a right-handed rectangular (Cartesian) system. This is the default coordinate system. Unless a different choice is made, GRID point locations, GRID point displacements, applied loads, and reactions will all be expressed using this system as reference.

The book is organized in three parts containing a total of twenty-eight examples. The first part deals with statics; the second part is devoted to dynamics; and the third part presents some useful modeling tricks. For convenience, the examples are all presented using a consistent system of units --there is no need to worry about conversions.

Each chapter presents a different example. The statement of the problem is followed by a detailed explanation of how the input file is set up. The results are then examined and discussed. All the relevant portions of output file, as well as the plots, are included with each chapter.

The possibilities are tremendous and the learning process need not be tedious. So, relax and enjoy the experience. In a few weeks you will be surprised by this new tool. Have fun and good luck!

---

[2] When using cylindrical coordinates (1,2,3) refers to (r, θ, z), and when using spherical coordinates (1,2,3) refers to (r, θ, ϕ).

PART 1

STATICS

1
Problem 1

1.1 Statement of the problem

Consider the simple truss shown in Figure 1.1. This is a two-dimensional structure made of steel. The geometry is specified in the figure. Assume a value of E (Elasticity modulus) equal to 3×10^7 psi. A force of 120 lb is applied horizontally at a point with coordinates $x = 60$ in and $y = 30$ in. Determine the displacements caused by this loading condition.

1.2 Cards introduced

Case Control Deck

DISPLACEMENT
ECHO
LOAD
TITLE

Bulk Data Deck

CROD
FORCE
GRID
MAT1
PROD

1.3 MSC/NASTRAN formulation

The structure can be modeled using ROD elements. The ROD is a finite element representing a straight bar of constant cross section. It can take compression (or tension) and torsion. In this problem the torsional capabilities of the ROD are not used. The ROD element does not support shear or bending.

Six GRID points and nine RODs are used to model the truss. Due to the nature of the problem, only two degrees of freedom per GRID point are needed to adequately model the behavior of the structure (translation in the x-direction and translation in the y-direction). Rotations about z are not required since there is no bending in any member of the truss. The finite element model is shown in Figure 1.2.

1.4 Input Data Deck

The input file used to solve this problem is shown in Exhibit 1.1. The organization of this file will be explained in detail.

The input file has three different sections: the Executive Control Deck, the Case Control Deck and the Bulk Data Deck.

1.4.1 Executive Control Deck

The first portion of the input file is the Executive Control Deck. It consists of all the cards or instructions (or, more properly, card images) located between the ID card (the first card) and the CEND card.

The ID card consists of the "word" ID followed by two alphanumeric sequences of characters separated by a comma.

The SOL card indicates the type of analysis to be performed. The number 24 corresponds to linear static analysis. Other types of analyses and their corresponding codes are listed in Appendix I.

The TIME card specifies the maximum number of CPU *minutes* (not seconds) that the job is allowed to run.

The CEND card marks the end of the Executive Control Deck. The format in this deck is free and an arbitrary number of blanks may be inserted on each line.

1.4.2 Case Control Deck

The second portion of the Input Data Deck is the Case Control Deck. This deck consists of all card images between the CEND card and the BEGIN BULK card. The main function of the Case Control Deck is to select certain cards of the Bulk Data Deck that specify particular features of the model or the solution technique to be used, to specify the loads acting on the structure under study, and to specify the desired type of output. The cards required to generate plots go at the end of this deck, but they are optional.

Consider the example in Exhibit 1.1. The TITLE card defines a heading that is printed at the top of each page of the output file.

The ECHO=BOTH card is a useful card for new users. It requests the Bulk Data Deck to be printed in the output file in two different ways: first, the same way it was typed in the input file, and second, sorted by alphabetical order.

The LOAD card specifies the load acting on the structure. It selects a "load" card, in this case a card with an ID equal to 123, from the Bulk Data Deck.

The DISPLACEMENT=ALL card is to ask for the displacement vector to be printed for all GRID points.

The dollar sign ($) is used to insert comments in the Case Control Deck. Whenever the $ appears in the first column of a card, that card is ignored. In this example, for clarity, a dollar signs have been used to create blank lines to separate the plot cards from the rest of the Case Control Deck.

The cards to create the plots are placed at the bottom of this deck. These are the cards that go between OUTPUT(PLOT) and BEGIN BULK.

The SET 22 INCLUDE ALL card specifies that the plots will include all the elements of the model.

The AXES Z, X, Y card specifies the way we are going to "look at the structure" (see Figure 1.3) in the plots. The choice R=Z, S=X, and T=Y states that the structure will be projected onto the x-y plane. This is consistent with the fact that this is a two-dimensional truss defined in the x-y plane. In other words, the plotting plane (S-T plane) is lined up with the x-y plane of the structure. For convenience the rotations (α, β, γ) are chosen to be 0. This is done with the VIEW 0.0, 0.0, .0 card.

The FIND card is used to request MSC/NASTRAN to determine the origin, the scale factor, and the vantage point for the plots using the information already specified by the user.

PLOT LABEL BOTH is used to plot the finite element model. Both the GRID points and the finite elements (in this case the ROD elements) are labeled. PLOT STATIC DEFORMATION 0 SET 22 is used to plot both the deformed and the undeformed structure. If the number 0 between DEFORMATION and SET is omitted only the deformed structure is plotted.

1.4.3 Bulk Data Deck

Finally, consider the Bulk Data Deck, which consists of the statements or card images between BEGIN BULK and ENDDATA. This is the portion of the input file that defines the finite element model. Typically, this deck constitutes the largest portion of the input file.

If we examine the input file used to solve this problem (Exhibit 1.1), we will see that comments can also be included in the Bulk Data Deck using the $ symbol. Whenever there is a dollar sign at the beginning of a card, that line is ignored. The cards in the Bulk Data Deck do not have to follow any particular sequence, although it is a good idea to organize them in a consistent fashion (for example, all the GRID cards together, all the PROD cards together, etc.).

The finite element model of the truss is shown in Figure 1.2. First, we define the GRID points (nodes) according to the layout shown in Figure 1.2 using GRID cards. Six GRID cards are required, one for each GRID point.[1] The third and the seventh field of these cards are blank. The blank in the third field indicates that the basic coordinate system (default) is employed to define the location of the GRID points. The blank in the seventh field indicates that the basic coordinate system is also used to specify the displacements once the problem is solved. Field number 8 can be used to constrain (eliminate) some degrees of freedom. Recall that to model this structure it is not necessary to include degrees of freedom 3, 4, 5, and 6 at each GRID point. In addition, due to the support conditions, degrees of freedom 1 and 2 (translation in the x- and y-direction) at GRID 1, and degree of freedom 2 (translation in the y-direction) at GRID 4 must be constrained.

The cards in the Bulk Data Deck can be entered into the computer using two different formats. The first alternative is the one illustrated with the GRID cards. Since each card of this deck has ten fields, commas can be used to separate the entries in each field. Note that an arbitrary number of blanks can be placed between two commas to specify that this field is indeed blank. Even two commas with no blanks in between will serve this purpose. This format is rather flexible. Notice also that 0.0, .0 or 0. are all valid choices to specify a zero. The connectivity of the finite elements of the model, in this case ROD elements, is specified with CROD cards. A different CROD card is used for each ROD element. The second field on this card is the identification (ID) number of the element. This must be an integer. The only restriction is that ID numbers must be unique. They do not have to be consecutive numbers. The third field points to a PROD card used to specify certain geometric properties of the RODs. Notice that the order in which the end points (GRID1 and GRID2) are specified is not important from a structural standpoint. G1=1 and G2=6 has the same effect as G1=6 and G2=1; that is, a ROD connecting GRIDs 1 and 6.

A second alternative to arrange the entries on any card of the Bulk Data Deck is the one demonstrated with the CROD cards. The first eight columns are employed to specify the first entry of the card; the second eight columns (columns 9 to 16) are used for the second entry; columns 17 to 24 are used

---

[1] Appendix II includes a detailed description of the Bulk Data Deck cards.

for the third entry, and so on. No commas are needed with this approach. This is an easy way to input the data if tabs can be used with the screen editor. Otherwise, it can be a little uncomfortable since you will have to count the spaces to make sure you are putting the right data in the right place. In any event, both approaches are equally valid and they can be mixed at will. You can use whichever one you prefer.

The PROD card specifies certain information regarding the geometry of the RODs. The second field of this card is the ID number, which has been referenced by a CROD card. The third field points to a MAT1 card (used to describe material properties). The fourth field gives the area of the cross section of the ROD. The RODs that have the same cross section can be described with the same PROD card. Not all the fields of this card need to be filled out. For example, torsion is not involved in this problem, so the value of J can be omitted in the fifth field.

Since the structure is made of only one material (steel), and therefore all the finite elements share the same material properties, one MAT1 card suffices to describe the material. Only E needs to be specified here.

Note that real numbers in the Bulk Data Deck can be expressed in several ways. For instance, 22.45, 0.2245+2, 2.245E+1, 2245.0-2, etc., are all valid choices to write the same number.

The force acting on this structure is a horizontal force of 120 lb applied at GRID 5. A FORCE card is used to describe it. The fourth field is blank indicating that the basic coordinate system (default) is used to specify the orientation of the force. The FORCE card is selected by the card LOAD = 123 (123 is an arbitrary ID number) in the Case Control Deck. If we did not have a LOAD card in the Case Control Deck this FORCE card would have been ignored.

1.5 Results

The output file corresponding to this problem is shown in Exhibit 1.2. First, notice that the first part of the output file shows the input file separated in its three different subdecks. Page 1 of the output file shows the Executive Control Deck.[2] Page 2 shows the Case Control Deck. Page 3 includes the Bulk Data Deck just the way it was typed in the input file and Page 4 shows the Bulk Data Deck sorted by alphabetical order. Comment cards are omitted in sorted format. The rest of the output file contains what are considered the actual "results".

---

[2] This page number refers to the pages in the output file, not in the book.

A number called EPSILON, which is very small in this case, appears at the bottom of page 5 of the output file. This number is always computed in linear static analyses (SOL 24). Its significance is discussed in the next example.

The only output requested was the displacement for all GRID points (DISPLACEMENT=ALL in the Case Control Deck). This output is shown on page 6 of the output file. As expected, the displacements for those degrees of freedom constrained using the eighth field of the GRID card are zero.

Two plots were generated by this run. Pages 5 and 8 of the output file show the options chosen to make these plots. The first plot (Figure 1.4) is produced by PLOT LABEL BOTH. It shows the finite element model including the ID numbers of both the RODs and the GRID points. The second plot (Figure 1.5), generated with PLOT STATIC DEFORMATION 0 SET 22 superimposes plots of the undeformed structure and the deformed structure.

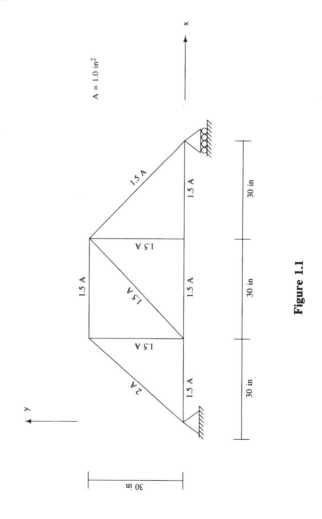

Figure 1.1

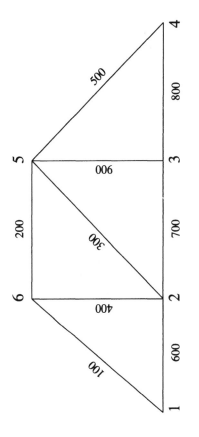

Figure 1.2

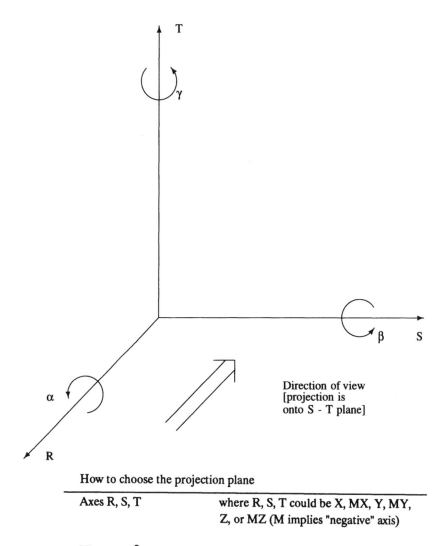

How to choose the projection plane

| Axes R, S, T | where R, S, T could be X, MX, Y, MY, Z, or MZ (M implies "negative" axis) |
| View γ, β, α | angles of rotation |

Figure 1.3

```
ID    PROBLEM,ONE
SOL   24
TIME  5
CEND
TITLE=  SIMPLE 2-D TRUSS; STATIC ANALYSIS
ECHO= BOTH
LOAD = 123
DISPLACEMENT= ALL
$
$  PLOTS
$
OUTPUT(PLOT)
SET 22  INCLUDE  ALL
AXES  Z,X,Y
VIEW  0.0, 0.0, .0
FIND
PLOT  LABEL  BOTH
PLOT  STATIC  DEFORMATION  0  SET  22
$
BEGIN BULK
$
$  GRID POINTS OF THE MODEL
$
GRID,1,  ,0.0, .0, 0., ,123456
GRID,2,  ,30.0,  .0, .0,,3456
GRID,3,, 60.0,  0.0, .0,,3456
GRID,4,  ,90.00, .0, .0,   ,23456
GRID,5,  ,60.00, 30.0,  .0,,3456
GRID,6,,30.0, 30.0,  .0,, 3456
$
```

Exhibit 1.1   Input file

```
$ NOW WE SPECIFY THE CONNECTIVITY OF THE ROD ELEMENTS
$
CROD    100      666      1      6
CROD    200      777      6      5
CROD    300      777      2      5
CROD    400      777      6      2
CROD    900      777      5      3
CROD    500      777      4      5
CROD    600      777      1      2
CROD    700      777      2      3
CROD    800      777      4      3
$
$ GEOMETRIC PROPERTIES OF THE RODS
$
PROD,666,999999,2.0
PROD   777      999999   1.5
$
$ MATERIAL PROPERTIES
$
MAT1,999999,30.0+6,
$
$ FORCE ACTING AT GRID 5
$
FORCE  123      5                    120.0   1.000   0.000   0.0
$
ENDDATA
```

Exhibit 1.1 (continued)   Input file

```
JUNE  6, 1988  MSC/NASTRAN 9/ 2/87    PAGE    1


N A S T R A N   E X E C U T I V E   C O N T R O L   D E C K   E C H O




    ID    PROBLEM,ONE
    SOL   24
    TIME  5
    CEND

    SIMPLE 2-D TRUSS; STATIC ANALYSIS
JUNE  6, 1988  MSC/NASTRAN 9/ 2/87    PAGE    2




C A S E    C O N T R O L   D E C K   E C H O
    CARD
    COUNT
     1      TITLE=  SIMPLE 2-D TRUSS; STATIC ANALYSIS
     2      ECHO= BOTH
     3      LOAD = 123
     4      DISPLACEMENT= ALL
     5      $
     6      $  PLOTS
     7      $
     8      OUTPUT(PLOT)
     9      SET 22  INCLUDE  ALL
    10      AXES  Z,X,Y
    11      VIEW  0.0, 0.0, .0
    12      FIND
    13      PLOT  LABEL  BOTH
    14      PLOT  STATIC  DEFORMATION  0  SET  22
    15      $
    16      BEGIN BULK
```

Exhibit 1.2 Output file

```
    SIMPLE 2-D TRUSS; STATIC ANALYSIS
JUNE  6, 1988  MSC/NASTRAN 9/ 2/87    PAGE     3


            I N P U T   B U L K   D A T A   D E C K   E C H O


  .   1 ..   2 ..   3 ..   4 ..   5 ..   6 ..   7 ..   8 ..
  $
  $  GRID POINTS OF THE MODEL
  $
  GRID,1,  ,0.0, .0, 0., ,123456
  GRID,2,  ,30.0,  .0, .0,,3456
  GRID,3,, 60.0, 0.0, .0,,3456
  GRID,4,  ,90.00, .0, .0,  ,23456
  GRID,5,  ,60.00, 30.0,  .0,,3456
  GRID,6,,30.0, 30.0,  .0,, 3456
  $
  $ NOW WE SPECIFY THE CONNECTIVITY OF THE ROD ELEMENTS
  $
  CROD    100    666     1      6
  CROD    200    777     6      5
  CROD    300    777     2      5
  CROD    400    777     6      2
  CROD    900    777     5      3
  CROD    500    777     4      5
  CROD    600    777     1      2
  CROD    700    777     2      3
  CROD    800    777     4      3
  $
  $ GEOMETRIC PROPERTIES OF THE RODS
  $
  PROD,666,999999,2.0
  PROD    777    999999  1.5
  $
  $ MATERIAL PROPERTIES
  $
  MAT1,999999,30.0+6,
  $
  $ FORCE ACTING AT GRID 5
  $
  FORCE   123    5                120.0  1.000  0.000  0.0
  $
  ENDDATA
  INPUT BULK DATA CARD COUNT =      36
```

Exhibit 1.2 (continued)   Output file

```
        SIMPLE 2-D TRUSS; STATIC ANALYSIS
   JUNE  6, 1988  MSC/NASTRAN 9/ 2/87   PAGE     4




                        S O R T E D    B U L K    D A T A    E C H O
   CARD
   COUNT  . 1  .. 2  .. 3  .. 4  .. 5  .. 6  .. 7  .. 8  ..
     1-   CROD   100    666    1     6
     2-   CROD   200    777    6     5
     3-   CROD   300    777    2     5
     4-   CROD   400    777    6     2
     5-   CROD   500    777    4     5
     6-   CROD   600    777    1     2
     7-   CROD   700    777    2     3
     8-   CROD   800    777    4     3
     9-   CROD   900    777    5     3
    10-   FORCE  123    5            120.0  1.000  0.000  0.0
    11-   GRID   1             0.0    .0    0.              123456
    12-   GRID   2             30.0   .0    .0              3456
    13-   GRID   3             60.0   0.0   .0              3456
    14-   GRID   4             90.00  .0    .0              23456
    15-   GRID   5             60.00  30.0  .0              3456
    16-   GRID   6             30.0   30.0  .0              3456
    17-   MAT1   999999 30.0+6
    18-   PROD   666    999999 2.0
    19-   PROD   777    999999 1.5
          ENDDATA
```

Exhibit 1.2 (continued)   Output file

```
SIMPLE 2-D TRUSS; STATIC ANALYSIS                          JUNE   6 , 1988   MSC/NASTRAN 9/ 2/87    PAGE     5

                              MESSAGES FROM THE PLOT MODULE


P L O T T E R    D A T A


    THE FOLLOWING PLOTS ARE FOR A NASTPLT PLOTTER


    PAPER SIZE = 20.0 X 20.0,   PAPER TYPE = VELLUM


    PEN 1 - SIZE 1, BLACK
    PEN 2 - SIZE 1, BLACK
    PEN 3 - SIZE 1, BLACK
    PEN 4 - SIZE 1, BLACK
```

Exhibit 1.2 (continued) Output file

E N G I N E E R I N G   D A T A   JUNE   6, 1988   MSC/NASTRAN 9/ 2/87   PAGE   5 (CONTINUED)

    ORTHOGRAPHIC PROJECTION
    ROTATIONS (DEGREES) - GAMMA =   0.00, BETA =   0.00, ALPHA =   0.00,   AXES = +Z,+X,+Y, SYMMETRIC
    SCALE (OBJECT-TO-PLOT SIZE) = 2.062222E-01

L I S T   O F   P L O T S

    PLOT   1   UNDEFORMED SHAPE

*** USER INFORMATION MESSAGE 5293 FOR DATA BLOCK KLL

LOAD SEQ. NO.   EPSILON   EXTERNAL WORK   EPSILONS LARGER THAN 0.001 ARE FLAGGED WITH ASTERISKS
      1   -4.9360868E-17   8.4150266E-03

Exhibit 1.2 (continued) Output file

SIMPLE 2-D TRUSS; STATIC ANALYSIS   JUNE 6, 1988 MSC/NASTRAN 9/ 2/87   PAGE   6

D I S P L A C E M E N T   V E C T O R

POINT ID.	TYPE	T1	T2	T3	R1	R2	R3
1	G	0.0	0.0	0.0	0.0	0.0	0.0
2	G	5.333333E-05	-3.034856E-05	0.0	0.0	0.0	0.0
3	G	8.000000E-05	-4.184095E-05	0.0	0.0	0.0	0.0
4	G	1.066667E-04	0.0	0.0	0.0	0.0	0.0
5	G	1.402504E-04	-4.184095E-05	0.0	0.0	0.0	0.0
6	G	1.135838E-04	-5.701523E-05	0.0	0.0	0.0	0.0

Exhibit 1.2 (continued)   Output file

SIMPLE 2-D TRUSS; STATIC ANALYSIS   JUNE 6, 1988 MSC/NASTRAN 9/ 2/87   PAGE   8

MESSAGES FROM THE PLOT MODULE

P L O T T E R   D A T A

THE FOLLOWING PLOTS ARE FOR A NASTPLT PLOTTER
PAPER SIZE = 20.0 X 20.0, PAPER TYPE = VELLUM

PEN 1 - SIZE 1, BLACK
PEN 2 - SIZE 1, BLACK
PEN 3 - SIZE 1, BLACK
PEN 4 - SIZE 1, BLACK

E N G I N E E R I N G   D A T A
ORTHOGRAPHIC PROJECTION
ROTATIONS (DEGREES) - GAMMA =   0.00, BETA =   0.00, ALPHA =   0.00, AXES = +Z,+X,+Y, SYMMETRIC
SCALE (OBJECT-TO-PLOT SIZE) = 1.757768E-01

L I S T   O F   P L O T S
PLOT   2   STATIC DEFORM.   1 - SUBCASE   123 - LOAD
* * * END OF JOB * * *

Exhibit 1.2 (continued) Output file

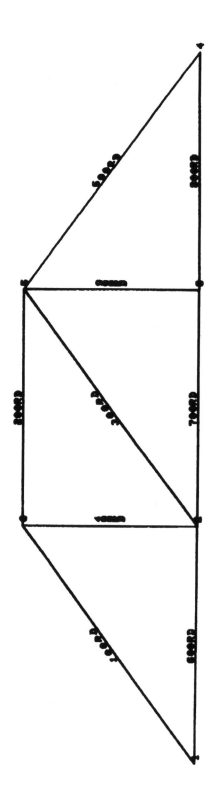

SIMPLE 2-D TRUSS   STATIC ANALYSIS

UNDEFORMED SHAPE

Figure 1.4 Finite element model of a plane truss using rod elements.

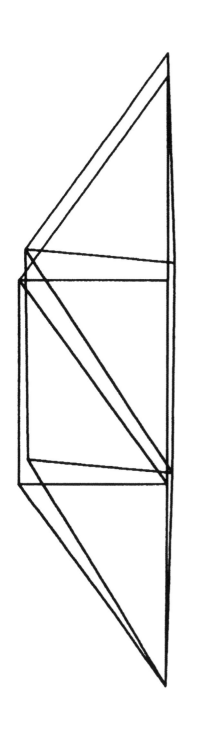

SIMPLE 2-D TRUSS STATIC ANALYSIS
STATIC DEFOR. SUBCASE 1 LOAD 123

Figure 1.5 Deformed and undeformed shape of the truss.

2
Problem 2

2.1 Statement of the problem

Consider the structure analyzed in Problem 1 (Figures 1.1 and 1.2). Assume that the displacements as well as the reactions, stresses, and internal element forces need to be computed. Therefore, additional data recovery requests need to be included in the Case Control Deck.

2.2 Cards introduced

Case Control Deck

ELFORCES
OLOAD
SPCFORCES
STRESS
SUBTITLE

Bulk Data Deck

None

2.3 MSC/NASTRAN formulation

The same model used in Problem 1 and depicted in Figure 1.2 is used. Thus, the Bulk Data Deck does not change.

2.4 Input Data Deck

The input file (Executive Control Deck, Case Control Deck, and Bulk Data Deck) is automatically printed at the beginning of the output file. Therefore, the input file will no longer be shown and reference will be made to the corresponding section of the output file for all future discussions.

Exhibit 2.1 shows the Executive Control Deck. Recall that comments can be included in this deck by using a dollar sign.

Consider the Case Control Deck shown in Exhibit 2.2. The purpose of the SUBTITLE card, which is similar to the TITLE card, is to include additional information describing the analysis. Both the TITLE and the SUBTITLE are printed on top of each page of the output file. The dollar sign can also be used to include comments in the same line an instruction is typed; whatever comes after the dollar sign is considered a comment. The OLOAD=ALL card requests that the external (applied) loads be printed. This is useful to verify that the applied loads are correctly entered. STRESS=ALL prints the stresses for each element of the model. ELFORCE=ALL requests that the internal forces in each element be printed. SPCFORCES=ALL requests the single point constraint forces for all supports. In other words, it asks for the reaction forces and moments at the GRID points where the structure is supported. These Case Control cards, requesting different types of recovery quantities, can be arranged in any order. A complete list of these cards is given in Appendix III. Notice that no plots requests are included in the Case Control Deck.

The Bulk Data Deck is shown in Exhibits 2.3 and 2.4.

2.5 Results

Exhibits 2.5 to 2.10 show selected portions of the output file with the results of the analysis.

First, examine EPSILON shown in Exhibit 2.5. This number must always be checked in linear static analysis (SOL 24). In this type of analysis a system of equations of the form

$$[K]\{x\} = \{f\} \tag{2.1}$$

is solved, where $[K]$ represents the stiffness matrix of the structure; $\{x\}$ represents the displacement vector (unknown); and $\{f\}$ represents the vector of the applied loads[1].

After MSC/NASTRAN performs a sequence of operations to solve this equation (Choleski's triangular decomposition method followed by the corresponding forward and backward substitution) a "solution" , say $\{x'\}$, is found. If $\{x\}$ in Eq. 2.1 is substituted by $\{x'\}$, generally

---

[1] [ ] denotes a matrix, { } denotes a column vector.

$$[K]\{x'\}-\{f\}=0 \qquad (2.2)$$

will not be obtained. In all probability the result will be

$$[K]\{x'\}-\{f\}=\{\delta\} \qquad (2.3)$$

where $\{\delta\}$ is the error due to computer roundoff in the "solution" $\{x'\}$. Obviously, $\{\delta\}$ should be very small to ensure that $\{x'\}$ is a good approximation to the true solution $\{x\}$. But how small should $\{\delta\}$ be? One way to evaluate the quality of the "solution" is the following. $\{\delta\}$ is an error vector that has dimensions of force. In a way, it can be considered to be the unbalanced force. Therefore, $<\delta>\{x'\}$ represents the work done by the unbalanced force[2]. Similarly, $<f>\{x'\}$ represents the work done by the external force. Therefore, if the work done by the unbalanced (error) force is small compared to the work done by the external force, the approximate "solution" $\{x'\}$ can be considered acceptable. Hence, a parameter EPSILON is defined such that

$$EPSILON = <\delta>\{x'\}/<f>\{x'\} \qquad (2.4)$$

A small EPSILON (a number smaller than about .00001 in absolute value) indicates a reasonably accurate solution. A large EPSILON indicates that the results are likely to be affected by severe roundoff error and are therefore unreliable. This means that the stiffness matrix of the structure is ill-conditioned (determinant close to 0).

An ill-conditioned stiffness matrix often occurs when the structure is not properly constrained or when there is a member that is significantly (several orders of magnitude) more stiff than neighboring members of the structure. It is necessary to determine the reason for a large EPSILON and to repeat the analysis after making the appropriate corrections.

Exhibit 2.6 shows the displacement at each GRID point. These displacements are identical to those obtained in Problem 1.

The output shown in Exhibit 2.7 is the result of OLOAD=ALL in the Case Control Deck. It indicates that the only applied load is a load of 120 lb at GRID 5 acting in the x-direction (T1). Exhibit 2.8 shows the reaction forces (single point constraint forces). This output is produced by SPCFORCES=ALL in the Case Control Deck. In this context, T1, T2, and T3 refer to the reaction forces in the x-, y-, and z-directions and R1, R2, and R3 refer to the moments about the x-, y-, and z-directions. Exhibit 2.9 shows the internal forces in each ROD element. This output is produced by ELFORCE=ALL in the Case Control Deck. Note that element 900 is not taking any force. Finally, since there is no torsion in any member of the truss,

---

[2] $< >$ denotes a row vector.

the value of torque is zero for all elements. Exhibit 2.10 shows the axial stresses in each ROD. This output is the result of STRESS = ALL in the Case Control Deck.

```
   N A S T R A N   E X E C U T I V E   C O N T R O L   D E C K   E.C H O

   ID    PROBLEM,TWO
   SOL   24    $  WE CAN ALSO PUT COMMENTS HERE IF WE WANT
   TIME  5     $  THIS IS THE MAX. CPU TIME IN MINUTES
   CEND
```

Exhibit 2.1   Executive Control Deck

```
   SIMPLE 2-D TRUSS; STATIC ANALYSIS
  SAME AS PROB. 1 BUT WITH MORE OUTPUT


               C A S E    C O N T R O L   D E C K   E C H O
 CARD
 COUNT
  1  TITLE=     SIMPLE 2-D TRUSS; STATIC ANALYSIS
  2  SUBTITLE=  SAME AS PROB. 1 BUT WITH MORE OUTPUT
  3  $
  4  ECHO= BOTH
  5  $
  6  OLOAD= ALL       $  TO PRINT THE EXTERNAL LOADS
  7  STRESS= ALL      $  REQUEST THE ELEMENT STRESS
  8  $
  9  LOAD = 123
 10  DISPLACEMENT= ALL
 11  $
 12  ELFORCE  = ALL  $  PRINT    ELEMENT FORCES
 13  SPCFORCES= ALL  $  PRINT REACTIONS (SINGLE POINT CONSTRAINT FORCES)
 14     $
 15        BEGIN BULK
```

Exhibit 2.2   Case Control Deck

```
                I N P U T   B U L K   D A T A   D E C K   E C H O

   .   1  ..   2  ..   3  ..   4  ..   5  ..   6  ..   7  ..   8  ..
   $
   $  GRID POINTS OF THE MODEL
   $
   GRID,1,  ,0.0, .0, 0., ,123456
   GRID,2,  ,30.0,  .0, .0,,3456
   GRID,3,, 60.0,  0.0, .0,,3456
   GRID,4,  ,90.00, .0, .0,   ,23456
   GRID,5,  ,60.00, 30.0,  .0,,3456
   GRID,6,,30.0, 30.0,  .0,, 3456
   $
   $ NOW WE SPECIFY THE CONNECTIVITY OF THE ROD ELEMENTS
   $
   CROD     100     666     1       6
   CROD     200     777     6       5
   CROD     300     777     2       5
   CROD     400     777     6       2
   CROD     900     777     5       3
   CROD     500     777     4       5
   CROD     600     777     1       2
   CROD     700     777     2       3
   CROD     800     777     4       3
   $
   $ GEOMETRIC PROPERTIES OF THE RODS
   $
   PROD,666,999999,2.0
   PROD    777     999999  1.5
   $
   $ MATERIAL PROPERTIES
   $
   MAT1,999999,30.0+6,
   $
   $ FORCE ACTING AT GRID 5
   $
   FORCE   123     5                 120.0   1.000   0.000   0.0
   $
   ENDDATA

   INPUT BULK DATA CARD COUNT =      36
```

Exhibit 2.3   Input Bulk Data Deck

```
        SIMPLE 2-D TRUSS; STATIC ANALYSIS
        SAME AS PROB. 1 BUT WITH MORE OUTPUT
                                JUNE 6, 1988 MSC/NASTRAN 9/ 2/87 PAGE  4


                            S O R T E D   B U L K   D A T A   E C H O
CARD
COUNT  .  1  ..  2  ..  3  ..  4  ..  5  ..  6  ..  7  ..  8
    1-  CROD   100   666    1     6
    2-  CROD   200   777    6     5
    3-  CROD   300   777    2     5
    4-  CROD   400   777    6     2
    5-  CROD   500   777    4     5
    6-  CROD   600   777    1     2
    7-  CROD   700   777    2     3
    8-  CROD   800   777    4     3
    9-  CROD   900   777    5     3
   10-  FORCE  123   5            120.0   1.000   0.000   0.0
   11-  GRID   1            0.0    .0     0.              123456
   12-  GRID   2            30.0   .0     .0              3456
   13-  GRID   3            60.0   0.0    .0              3456
   14-  GRID   4            90.00  .0     .0              23456
   15-  GRID   5            60.00  30.0   .0              3456
   16-  GRID   6            30.0   30.0   .0              3456
   17-  MAT1   999999  30.0+6
   18-  PROD   666   999999  2.0
   19-  PROD   777   999999  1.5
        ENDDATA

        TOTAL COUNT=   20
```

Exhibit 2.4   Bulk Data Deck sorted by alphabetical order

```
*** USER INFORMATION MESSAGE 5293 FOR DATA BLOCK KLL

LOAD SEQ. NO.      EPSILON          EXTERNAL WORK      EPSILONS LARGER THAN 0.001 ARE FLAGGED WITH ASTERISKS
     1          -4.9360868E-17     8.4150266E-03
```

Exhibit 2.5 Epsilon

D I S P L A C E M E N T   V E C T O R

POINT ID.	TYPE	T1	T2	T3	R1	R2	R3
1	G	0.0	0.0	0.0	0.0	0.0	0.0
2	G	5.333333E-05	-3.034856E-05	0.0	0.0	0.0	0.0
3	G	8.000000E-05	-4.184095E-05	0.0	0.0	0.0	0.0
4	G	1.066667E-04	0.0	0.0	0.0	0.0	0.0
5	G	1.402504E-04	-4.184095E-05	0.0	0.0	0.0	0.0
6	G	1.135838E-04	-5.701523E-05	0.0	0.0	0.0	0.0

Exhibit 2.6 Displacement vector

```
SIMPLE 2-D TRUSS; STATIC ANALYSIS              JUNE  6, 1988  MSC/NASTRAN 9/ 2/87    PAGE   6
SAME AS PROB. 1 BUT WITH MORE OUTPUT

                                    L O A D   V E C T O R

POINT ID.  TYPE        T1            T2         T3        R1        R2        R3
    5        G     1.200000E+02     0.0        0.0       0.0       0.0       0.0
```

Exhibit 2.7 Load vector

```
SIMPLE 2-D TRUSS; STATIC ANALYSIS              JUNE  6, 1988  MSC/NASTRAN 9/ 2/87    PAGE   7
SAME AS PROB. 1 BUT WITH MORE OUTPUT

                  F O R C E S   O F   S I N G L E - P O I N T   C O N S T R A I N T

POINT ID.  TYPE        T1            T2             T3        R1        R2        R3
    1        G     -1.200000E+02  -4.000000E+01    0.0       0.0       0.0       0.0
    4        G      0.0            4.000000E+01    0.0       0.0       0.0       0.0
```

Exhibit 2.8 Single point constraint forces

FORCES IN ROD ELEMENTS (CROD)

ELEMENT ID.	AXIAL FORCE	TORQUE	ELEMENT ID.	AXIAL FORCE	TORQUE
100	5.656854E+01	0.0	200	4.000000E+01	0.0
300	5.656854E+01	0.0	400	-4.000000E+01	0.0
500	-5.656854E+01	0.0	600	8.000000E+01	0.0
700	4.000000E+01	0.0	800	4.000000E+01	0.0
900	0.0	0.0			

Exhibit 2.9 Forces in ROD elements

STRESSES IN ROD ELEMENTS (CROD)

ELEMENT ID.	AXIAL STRESS	SAFETY MARGIN	TORSIONAL STRESS	SAFETY MARGIN	ELEMENT ID.	AXIAL STRESS	SAFETY MARGIN	TORSIONAL STRESS	SAFETY MARGIN
100	2.828427E+01		0.0	0.0	200	2.666667E+01		0.0	0.0
300	3.771236E+01		0.0	0.0	400	-2.666667E+01		0.0	0.0
500	-3.771236E+01		0.0	0.0	600	5.333333E+01		0.0	0.0
700	2.666667E+01		0.0	0.0	800	2.666667E+01		0.0	0.0
900	0.0		0.0	0.0					

Exhibit 2.10 Stresses in ROD elements

3
Problem 3

3.1 Statement of the problem

Consider the truss described in Problem 1 (Figure 1.1) subjected to an enforced displacement of 0.01 inches in the x-direction, applied at GRID point 4. Determine how the structure reacts.

3.2 Cards introduced

Case Control Deck   SPC

Bulk Data Deck   SPC
                                 SPC1

3.3 MSC/NASTRAN formulation

The same model used in Problems 1 and 2 and shown in Figure 1.2 is employed. The use of the SPC and SPC1 cards to constrain selected degrees of freedom is demonstrated. This is an alternative to the use of the eighth field of the GRID card. In addition, a SPC card in the Bulk Data Deck is used to enforce the displacement at GRID 4.

3.4 Input Data Deck

The Executive Control Deck (Exhibit 3.1) is similar to those used in the previous examples.

Consider the Case Control Deck shown in Exhibit 3.2. Notice a new card in this deck, the SPC=6767 card (card 8). This card is used to select the SPC and SPC1 cards in the Bulk Data Deck.

The statements to generate the plots are similar to those employed in Problem 1 (see cards 17 and 18). Both statements request the deformed structure to be plotted; the only difference between the first plot request and the second is the number 0 between DEFORMATION and SET, which is missing in the second statement. If the 0 is included the deformed and the undeformed structure are plotted. If the 0 is omitted only the deformed structure is plotted.

The Bulk Data Deck (Exhibits 3.3 and 3.4) is quite similar to that of Problems 1 and 2. The FORCE card has been deleted (no external forces are being applied) and two new cards have been incorporated: SPC and SPC1.

Consider first the constraints that need to be applied. The eighth field of the GRID card is used to constrain the appropriate translational degrees of freedom (1, 2, or 3) at the corresponding GRID points. The rotational degrees of freedom (4, 5, and 6) are constrained using a SPC1 card. This card, with an ID equal to 6767, is selected by the SPC=6767 card in the Case Control Deck.

The SPC=6767 card in the Case Control Deck also points to the SPC card of the Bulk Data Deck. This card specifies that GRID point 4 is subjected to an enforced displacement of .01 inches in the x-direction.

3.5 Results

EPSILON (shown in Exhibit 3.5) is small, which means that the problem is well conditioned. This indicates that the results are reliable, at least from a numerical standpoint.

Exhibit 3.6 shows the displacement vector. GRID 4, as expected, exhibits a displacement of exactly .01 inches in the x-direction. Reactions (SPC forces) and stresses are presented in Exhibits 3.7 and 3.8. Notice that the OLOAD=ALL card (card number 5 in the Case Control Deck) does not generate any output in this case. This is due to the fact that no external force or moment loads are applied.

Two plots are generated. The first plot (Figure 3.1) shows both the undeformed shape and the deformed shape of the structure. The second plot (Figure 3.2) shows only the deformed shape of the truss.

```
N A S T R A N   E X E C U T I V E   C O N T R O L   D E C K   E C H O


    ID    PROBLEM,NUMBER-3
    SOL   24
    TIME  5
    CEND
```

Exhibit 3.1   Executive Control Deck

```
              C A S E    C O N T R O L   D E C K   E C H O


    CARD
    COUNT
      1       TITLE=      SIMPLE 2-D TRUSS; STATIC ANALYSIS
      2       SUBTITLE=   SAME TRUSS OF PROB. 1, (ENFORCED DISPLACEMENT)
      3       ECHO= BOTH
      4       DISPLACEMENT= ALL
      5       OLOAD=  ALL
      6       STRESS=  ALL
      7       SPCFORCES = ALL
      8       SPC  =6767
      9       $
     10       $  PLOTS
     11       $
     12       OUTPUT(PLOT)
     13       SET  44  INCLUDE  ALL
     14       AXES    Z,  X, Y
     15       VIEW   .0, .0, .0
     16       FIND
     17       PLOT  STATIC  DEFORMATION  0  SET   44
     18       PLOT  STATIC  DEFORMATION SET 44
     19       $
     20       BEGIN BULK
```

Exhibit 3.2   Case Control Deck

```
           I N P U T   B U L K   D A T A   D E C K   E C H O

   .   1  ..   2  ..   3  ..   4  ..   5  ..   6  ..   7  ..   8  ..   9
   $
   $  ENFORCE A DISPLACEMENT OF .01 INCHES AT GRID 4 (X-DIREC)
   $
   SPC     6767    4       1         0.01
   $
   $  NOW WE CONSTRAIN THE ROTATIONS AT THE GRID POINTS USING THE SPC1
   $  CARD (INSTEAD OF PUTTING 456 IN THE 8TH FIELD OF THE GRID CARD)
   $
   SPC1    6767    456     1         THRU    6
   $
   $  GRID POINTS OF THE MODEL
   $
   GRID,1, ,0.0, .0, 0., ,123
   GRID,2, ,30.0, .0, .0,,3
   GRID,3,, 60.0, 0.0, .0,,3
   GRID,4, ,90.00, .0, .0,   ,23
   GRID,5, ,60.00, 30.0, .0,,3
   GRID,6,,30.0, 30.0, .0,, 3
   $
   $ NOW WE SPECIFY THE CONNECTIVITY OF THE ROD ELEMENTS
   $
   CROD    100     666     1       6
   CROD    200     777     6       5
   CROD    300     777     2       5
   CROD    400     777     6       2
   CROD    900     777     5       3
   CROD    500     777     4       5
   CROD    600     777     1       2
   CROD    700     777     2       3
   CROD    800     777     4       3
   $
   $ GEOMETRIC PROPERTIES OF THE RODS
   $
   PROD,666,999999,2.0
   PROD    777     999999 1.5
   $
   $ MATERIAL PROPERTIES
   $
   MAT1,999999,30.0+6,
   ENDDATA
```

Exhibit 3.3 Input Bulk Data Deck

```
                    S O R T E D   B U L K   D A T A   E C H O
CARD
COUNT       .  1  ..  2  ..  3  ..  4  ..  5  ..  6  ..  7  ..  8
   1-    CROD    100    666    1      6
   2-    CROD    200    777    6      5
   3-    CROD    300    777    2      5
   4-    CROD    400    777    6      2
   5-    CROD    500    777    4      5
   6-    CROD    600    777    1      2
   7-    CROD    700    777    2      3
   8-    CROD    800    777    4      3
   9-    CROD    900    777    5      3
  10-    GRID    1             0.0    .0     0.                    123
  11-    GRID    2             30.0   .0     .0                    3
  12-    GRID    3             60.0   0.0    .0                    3
  13-    GRID    4             90.00  .0     .0                    23
  14-    GRID    5             60.00  30.0   .0                    3
  15-    GRID    6             30.0   30.0   .0                    3
  16-    MAT1    999999 30.0+6
  17-    PROD    666    999999 2.0
  18-    PROD    777    999999 1.5
  19-    SPC     6767   4      1      0.01
  20-    SPC1    6767   456    1      THRU   6
         ENDDATA

                    TOTAL COUNT=    21
```

Exhibit 3.4   Bulk Data Deck sorted by alphabetical order

```
*** USER INFORMATION MESSAGE 5293 FOR DATA BLOCK KLL

  LOAD SEQ. NO.        EPSILON          EXTERNAL WORK        EPSILONS LARGER THAN 0.001 ARE FLAGGED WITH
                                                            ASTERISKS
       1            -7.0161923E-18       7.6516502E+01
```

Exhibit 3.5 Epsilon

```
SIMPLE 2-D TRUSS; STATIC ANALYSIS                        JUNE  6, 1988   MSC/NASTRAN 9/ 2/87   PAGE   5
SAME TRUSS OF PROB. 1, (ENFORCED DISPLACEMENT)

                                   D I S P L A C E M E N T   V E C T O R
```

POINT ID.	TYPE	T1	T2	T3	R1	R2	R3
1	G	0.0	0.0	0.0	0.0	0.0	0.0
2	G	3.333333E-03	-4.444445E-03	0.0	0.0	0.0	0.0
3	G	6.666666E-03	-5.555555E-03	0.0	0.0	0.0	0.0
4	G	1.000000E-02	0.0	0.0	0.0	0.0	0.0
5	G	4.444445E-03	-5.555555E-03	0.0	0.0	0.0	0.0
6	G	4.444445E-03	-4.444445E-03	0.0	0.0	0.0	0.0

Exhibit 3.6 Displacement vector

```
SIMPLE 2-D TRUSS; STATIC ANALYSIS                            JUNE   6, 1988  MSC/NASTRAN 9/ 2/87   PAGE    6
SAME TRUSS OF PROB. 1, (ENFORCED DISPLACEMENT)

                      F O R C E S   O F   S I N G L E - P O I N T   C O N S T R A I N T

POINT ID.   TYPE        T1              T2             T3          R1          R2          R3
    1         G    -5.000000E+03       0.0            0.0         0.0         0.0         0.0
    4         G     5.000000E+03   -1.136868E-13      0.0         0.0         0.0         0.0
```

Exhibit 3.7 Single point constraint forces

```
SIMPLE 2-D TRUSS; STATIC ANALYSIS                            JUNE   6, 1988  MSC/NASTRAN 9/ 2/87   PAGE    8
SAME TRUSS OF PROB. 1, (ENFORCED DISPLACEMENT)

                          S T R E S S E S   I N   R O D   E L E M E N T S   ( C R O D )

ELEMENT      AXIAL      TORSIONAL    SAFETY       ELEMENT      AXIAL       SAFETY     TORSIONAL    SAFETY
  ID.        STRESS      STRESS      MARGIN         ID.        STRESS      MARGIN      STRESS      MARGIN
  100         0.0         0.0                       200         0.0                     0.0
  300     -1.136868E-13   0.0                       400     1.136868E-13                0.0
  500      2.273737E-13   0.0                       600     3.333333E+03                0.0
  700      3.333333E+03   0.0                       800     3.333333E+03                0.0
  900         0.0         0.0
```

Exhibit 3.8 Stresses in ROD elements

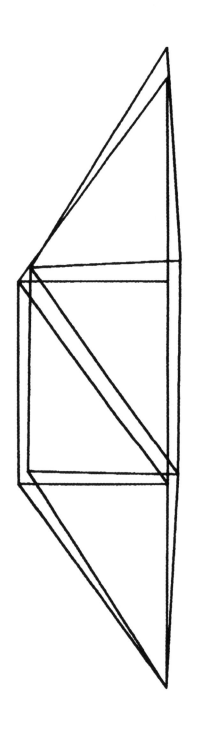

Figure 3.1 Deformed and undeformed shape of the truss.

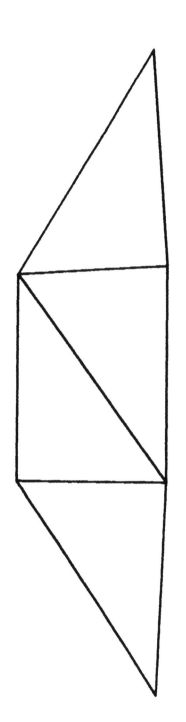

STATIC DEFOR. SUBCASE 1   LOAD 0

Figure 3.2 Deformed shape of the truss.

4

Problem 4

4.1 Statement of the problem

Analyze the truss described in Problem 1 (Figures 1.1 and 1.2) subjected to two different loading conditions. First, apply a thermal load by raising the temperature of the six GRIDs of the structure as follows:

$$T_1 = T_2 = T_3 = T_4 = 60\,°F \tag{4.1}$$

and

$$T_5 = T_6 = 200\,°F \tag{4.2}$$

Use a thermal expansion coefficient (α) of 6.3×10^{-6} in/in $°F$ and assume that the reference temperature (the temperature at which the structure is free of thermal stresses) is $20\,°F$. Second, determine the reaction of the structure if ROD number 100 is subjected to an axial elongation of .05 in.

4.2 Cards introduced

Case Control Deck   DEFORM
                                           LABEL
                                           SUBCASE
                                           TEMP

Bulk Data Deck   DEFORM
                                           TEMP

4.3 MSC/NASTRAN formulation

The same model as in Problems 1 through 3 is used (Figure 1.2). A TEMP card is employed to specify the temperatures at the GRID points and a DEFORM card is used to specify the deformation (elongation) applied to ROD 100. The analysis is carried out in one run even though two different loading conditions are specified. The SUBCASE card in the Case Control Deck allows the user to define several loading conditions for the same model.

4.4 Input Data Deck

The Executive Control Deck, shown in Exhibit 4.1, is similar to those of previous examples. SOL 24 is selected since this is a linear static analysis.

Two subcases are specified in the Case Control Deck (Exhibit 4.2) using the SUBCASE card. The first subcase is the thermal load subcase; the second is the enforced deformation subcase. The LABEL card is used to give a particular title to each subcase. The STRESS = ALL and DISPLACEMENT = ALL cards are placed above the subcase level, thereby affecting both subcases (displacements and stresses are printed for both). The TEMP = 333 card, which only affects subcase 1, selects the TEMP cards in the Bulk Data Deck. The DEFORM = 886622 card, which corresponds to the second subcase, points to a DEFORM card in the Bulk Data Deck. Note also that the SPC forces (reactions) are requested to be printed in the second subcase. The undeformed and the deformed structure are plotted for both subcases. Despite the fact that there are two subcases, only one PLOT statement is required.

Exhibits 4.3 and 4.4 show the Bulk Data Deck. The temperatures at the grid points are specified with TEMP cards. The TEMP cards are selected in the Case Control Deck by a TEMP card with the appropriate ID number (333 in this case). The DEFORM card whose ID is 886622 (consistent with the DEFORM = 886622 card in the Case Control Deck) specifies that ROD element 100 is subjected to an elongation of .05 inches.

The thermal expansion coefficient of the material and the reference temperature have been included in the seventh and eighth fields of the MAT1 card (α and T_{ref} must be consistent with the units employed in the TEMP cards). In this case the reference temperature is 20°F. This means that the variations in temperature are evaluated using 20°F as a baseline. For example, GRID point 6 is subjected to a change in temperature equal to 200-20 = 180°F. These changes in temperature ultimately determine the thermal stresses in the structure; the ambient value of the temperature is, to a certain extent, not important.

4.5 Results

The two EPSILONs shown in Exhibit 4.5 (one for each subcase) are acceptable.

Displacements and stresses are shown for both subcases (Exhibits 4.6 and 4.8) since DISPLACEMENT=ALL and STRESS=ALL cards were placed above the subcase level. The reaction forces (forces of single point constraint) are printed only for the second subcase (Exhibit 4.7), since the SPCFORCES=ALL card was placed after the SUBCASE 2 card in the Case Control Deck.

The text specified in the LABEL card of the Case Control Deck appears at the left of the page where output for that subcase is printed. Two plots are generated (Figures 4.1 and 4.2), one for each subcase.

4.6 Additional comments

In this example two problems were solve simultaneously, i.e. in one computer run. Two separate jobs could have been run, one for the thermal load case and a second for the enforced deformation case. This approach, however, would have been inefficient. The main advantage of solving for several load conditions (several subcases) in one run is, first, that the stiffness matrix of the structure is generated only once; and second, this stiffness matrix is decomposed only once (Choleski's decomposition) when solving the linear system $[K]\{x\} = \{f\}$. This saves considerable CPU time in large size problems.

```
N A S T R A N    E X E C U T I V E    C O N T R O L    D E C K    E C H O

    ID    PROBLEM,FOUR
    SOL    24
    TIME   5
    CEND
```

Exhibit 4.1   Executive Control Deck

```
                C A S E    C O N T R O L    D E C K    E C H O
        CARD
        COUNT
          1        TITLE= SIMPLE 2-D TRUSS; STATIC ANALYSIS
          2        ECHO= BOTH
          3        SUBTITLE= THERMAL LOADS AND ENFORCED DEFORMATIONS
          4        $
          5        STRESS= ALL
          6        DISPLACEMENT= ALL
          7        $
          8        SUBCASE  1
          9        LABEL=  THERMAL LOAD
         10        TEMP= 333  $ TO SPECIFY THE TEMPERATURE  AT EACH NODE
         11        $
         12        SUBCASE  2
         13        LABEL= ENFORCED DEFORMATION
         14        SPCFORCES=   ALL
         15        DEFORM= 886622
         16        $
         17        $ NOW  THE PLOTS
         18        $
         19        OUTPUT(PLOT)
         20        SET   22 INCLUDE  ALL
         21        AXES   Z,X,Y
         22        VIEW  .0,.0,.0
         23        FIND
         24        PLOT  STATIC DEFORMATION 0  SET  22
         25        BEGIN BULK
```

Exhibit 4.2   Case Control Deck

```
                  I N P U T   B U L K   D A T A   D E C K   E C H O
.   1 ..   2 ..   3 ..   4 ..   5 ..   6 ..   7 ..   8 ..   9
$
$ SPECIFY TEMPERATURE AT GRID POINTS
$
TEMP    333     1       60.00   2       60.0    3       60.0
TEMP,333,4,  60.0
TEMP    333     5       200.00  6       200.00
$
$ DEFORMATION ENFORCED
$
DEFORM,886622,100,  .05
$
$  GRID POINTS OF THE MODEL
$
GRID,1,  ,0.0, .0, 0., ,123456
GRID,2,  ,30.0,  .0, .0,,3456
GRID,3,, 60.0,  0.0, .0,,3456
GRID,4,  ,90.00, .0, .0,   ,23456
GRID,5,  ,60.00, 30.0,  .0,,3456
GRID,6,,30.0, 30.0,  .0,, 3456
$
$ NOW WE SPECIFY THE CONNECTIVITY OF THE ROD ELEMENTS
$
CROD    100     666     1       6
CROD    200     777     6       5
CROD    300     777     2       5
CROD    400     777     6       2
CROD    900     777     5       3
CROD    500     777     4       5
CROD    600     777     1       2
CROD    700     777     2       3
CROD    800     777     4       3
$
$ GEOMETRIC PROPERTIES OF THE RODS
$
PROD,666,999999,2.0
PROD    777     999999  1.5
$
$ MATERIAL PROPERTIES, ADD THERMAL EXP. COEFF. AND REFERENCE TEMP.
$
MAT1,999999,30.0+6,  ,  ,  ,6.3-6, 20.000
ENDDATA
```

Exhibit 4.3   Input Bulk Data Deck

```
                          S O R T E D   B U L K   D A T A   E C H O
CARD
COUNT    .  1  ..  2  ..  3  ..  4  ..  5  ..  6  ..  7  ..  8
  1-   CROD    100    666    1      6
  2-   CROD    200    777    6      5
  3-   CROD    300    777    2      5
  4-   CROD    400    777    6      2
  5-   CROD    500    777    4      5
  6-   CROD    600    777    1      2
  7-   CROD    700    777    2      3
  8-   CROD    800    777    4      3
  9-   CROD    900    777    5      3
 10-   DEFORM  886622 100    .05
 11-   GRID    1             0.0    .0     0.              123456
 12-   GRID    2             30.0   .0     .0              3456
 13-   GRID    3             60.0   0.0    .0              3456
 14-   GRID    4             90.00  .0     .0              23456
 15-   GRID    5             60.00  30.0   .0              3456
 16-   GRID    6             30.0   30.0   .0              3456
 17-   MAT1    999999 30.0+6               6.3-6           20.000
 18-   PROD    666    999999 2.0
 19-   PROD    777    999999 1.5
 20-   TEMP    333    1      60.00  2      60.0   3        60.0
 21-   TEMP    333    4      60.0
 22-   TEMP    333    5      200.00 6      200.00
       ENDDATA

       TOTAL COUNT=    23
```

Exhibit 4.4   Bulk Data Deck sorted by alphabetical order

```
*** USER INFORMATION MESSAGE 5293 FOR DATA BLOCK KLL

   LOAD SEQ. NO.          EPSILON              EXTERNAL WORK
          1          5.9928329E-17            3.1730952E+03
          2         -3.0316490E-17            1.7677670E+03
```

Exhibit 4.5   Epsilons

THERMAL LOAD   SUBCASE 1

D I S P L A C E M E N T   V E C T O R

POINT ID.	TYPE	T1	T2	T3	R1	R2	R3
1	G	0.0	0.0	0.0	0.0	0.0	0.0
2	G	7.560000E-03	2.646000E-02	0.0	0.0	0.0	0.0
3	G	1.512000E-02	2.646000E-02	0.0	0.0	0.0	0.0
4	G	2.268000E-02	0.0	0.0	0.0	0.0	0.0
5	G	2.835000E-02	4.725000E-02	0.0	0.0	0.0	0.0
6	G	-5.670000E-03	4.725000E-02	0.0	0.0	0.0	0.0

ENFORCED DEFORMATION   SUBCASE 2

D I S P L A C E M E N T   V E C T O R

POINT ID.	TYPE	T1	T2	T3	R1	R2	R3
1	G	0.0	0.0	0.0	0.0	0.0	0.0
2	G	-2.168404E-19	4.714045E-02	0.0	0.0	0.0	0.0
3	G	-2.710505E-19	2.357023E-02	0.0	0.0	0.0	0.0
4	G	-3.252607E-19	0.0	0.0	0.0	0.0	0.0
5	G	2.357023E-02	2.357023E-02	0.0	0.0	0.0	0.0
6	G	2.357023E-02	4.714045E-02	0.0	0.0	0.0	0.0

Exhibit 4.6 Displacement vectors

F O R C E S   O F   S I N G L E - P O I N T   C O N S T R A I N T

POINT ID.	TYPE	T1	T2	T3	R1	R2	R3
4	G	0.0	-2.273737E-13	0.0	0.0	0.0	0.0

Exhibit 4.7   Single point constraint forces for Subcase 2

S T R E S S E S   I N   R O D   E L E M E N T S   (C R O D)

THERMAL LOAD   SUBCASE   1

ELEMENT ID.	AXIAL STRESS	SAFETY MARGIN	TORSIONAL STRESS	SAFETY MARGIN	ELEMENT ID.	AXIAL STRESS	SAFETY MARGIN	TORSIONAL STRESS	SAFETY MARGIN
100	1.818989E-12		0.0		200	9.094947E-13		0.0	
300	1.364242E-12		0.0		400	-9.094947E-13		0.0	
500	1.364242E-12		0.0		600	-4.547474E-13		0.0	
700	-9.094947E-13		0.0		800	-9.094947E-13		0.0	
900	4.547474E-13		0.0						

ENFORCED DEFORMATION   SUBCASE   2

ELEMENT ID.	AXIAL STRESS	SAFETY MARGIN	TORSIONAL STRESS	SAFETY MARGIN	ELEMENT ID.	AXIAL STRESS	SAFETY MARGIN	TORSIONAL STRESS	SAFETY MARGIN
100	0.0		0.0		200	0.0		0.0	
300	-4.547474E-13		0.0		400	9.094947E-13		0.0	
500	2.921170E-13		0.0		600	-2.168404E-13		0.0	
700	-5.421011E-14		0.0		800	-5.421011E-14		0.0	
900	0.0		0.0						

Exhibit 4.8   Stresses in ROD elements

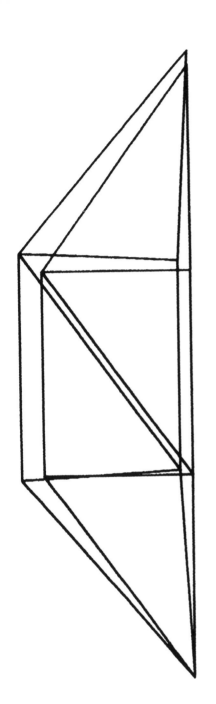

Figure 4.1 Deformed and undeformed shape of the truss (thermal load).

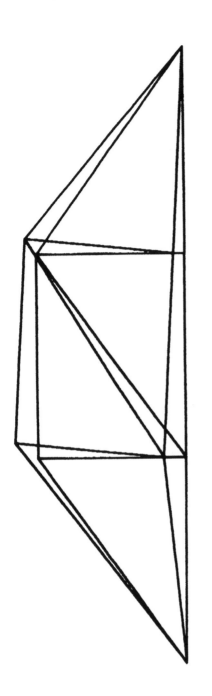

Figure 4.2 Deformed and undeformed shape of the truss
(enforced deformation).

5
Problem 5

5.1 Statement of the problem

Consider the truss described in Problem 1 (Figure 1.1). Determine the structure's reaction to the following loads:

1   A force of 333 lb applied at GRID 6 in the negative x-direction;

2   A force of 70 lb applied at GRID 3 in the y-direction and a force of 200 lb applied at GRID 6 in the y-direction;

3   The forces defined in (1) and (2) acting simultaneously;

4   The force defined in (1) multiplied by 0.5 and the forces defined in (2) multiplied by 1.66 acting simultaneously; and

5   The force defined in (1) multiplied by 0.4 and the forces defined in (2) multiplied by -2.0 acting simultaneously.

5.2 Cards Introduced

Case Control Deck   SET
                           SUBCOM
                           SUBSEQ

Bulk Data Deck   None

5.3 MSC/NASTRAN formulation

The model is shown in Figure 1.2. This problem demonstrates the use of the SUBCOM and SUBSEQ cards. These cards are used to define new loading conditions based on linear combinations of previously defined subcases. In this example, the first two loading conditions are defined as separate subcases. The remaining loading configurations are defined as linear combinations of the first two subcases. Linear combinations of previously defined subcases are possible because the structure is linear and the superposition principle holds.

5.4 Input Data Deck

Exhibit 5.1 shows the Executive Control Deck.

The Case Control Deck (Exhibit 5.2) introduces a new card, the SET card. This card is used to define a group of grids or elements. For example, SET 111 = 4, 2 specifies that SET 111 consists of two GRID points, 4 and 2; SET 55 = 400, 500 specifies that SET 55 consists of two finite elements, ROD 400 and ROD 500. The SET card can be used to request that quantities be printed only for some selected GRIDs or elements. For instance, DISPLACEMENT = 111, requests the displacements to be printed only for GRIDs 2 and 4. Analogously, ELFORCES = 55, requests the element forces to be output only for RODs 400 and 500.

The Case Control Deck includes two subcases, SUBCASE 1 and SUBCASE 2. The first subcase corresponds to the first loading condition and the second subcase defines the second loading condition. The SUBCOM and SUBSEQ cards at the end of the Case Control Deck define loading conditions based on linear combinations of SUBCASEs 1 and 2. The SUBCOM card labels the combination of loads. The SUBSEQ card specifies the factors used to combine the subcases. SUBCOM 450, for example, consists of the loads acting in the first subcase multiplied by a factor of 0.5 plus the loads acting in the second subcase multiplied by a factor of 1.66. The resulting set of forces is then applied to the structure.

The number of factors that appears in the SUBSEQ card *must* be equal to the number of subcases defined. If a certain subcase does not contribute to a certain combination, a 0.0 factor must be included in the appropriate place of the SUBSEQ card. *A blank is not acceptable* for this purpose. Notice also that the ID numbers of the SUBCOM cards are arbitrary integers that serve only to identify the combination. SUBCOM IDs need not be consecutive numbers (but must be increasing).

In this case the Bulk Data Deck was sorted by alphabetical order only (Exhibit 5.3). This is because the ECHO = BOTH card was omitted in the Case Control Deck.

Finally, note that the LOAD = 200 card in the second subcase of the Case Control Deck points to two FORCE cards in the Bulk Data Deck. Both FORCE cards are considered in the analysis, since they both have IDs of 200.

5.5 Results

Since there are two subcases, two EPSILONs are printed in the output file (Exhibit 5.4). Both are small, indicating that the problem is well conditioned.

Exhibit 5.5 presents the displacement vector for subcase 1 only at GRIDs 2 and 4, consistent with the request in the Case Control Deck. The external load vector is printed for the two subcases and the three combinations (Exhibit 5.6). The same happens with the stresses, which are printed in all five cases for all elements (Exhibit 5.8). The element forces, however, are printed only in the second subcase and for RODs 400 and 500, consistent with the request made in the Case Control Deck (Exhibit 5.7).

```
N A S T R A N   E X E C U T I V E   C O N T R O L   D E C K   E C H O

     ID    P5,P5
     SOL        24
     TIME    5
     CEND
```

Exhibit 5.1   Executive Control Deck

```
                    C A S E    C O N T R O L    D E C K    E C H O
CARD
COUNT
  1      TITLE=    SIMPLE 2-D TRUSS; STATIC ANALYSIS
  2      SUBTITLE= WE INTRODUCE NOW SEVERAL COMBINATIONS  OF CASES
  3      $
  4      STRESS=ALL
  5      SET   111=   4, 2
  6      SET 55=  400 , 500
  7      OLOAD =  ALL
  8      $
  9      SUBCASE   1
 10      LABEL =   HORIZONTAL   LOADS
 11               LOAD =100
 12               DISPLACEMENT =111
 13      $
 14      SUBCASE 2
 15      LABEL  =  VERTICAL   LOADS
 16               LOAD = 200
 17               ELFORCES=   55
 18      $
 19      $$$$$$$$$  COMBINATIONS $$$$$$$$$$$$
 20      $
 21        SUBCOM  400
 22        SUBSEQ=  1.0, 1.000
 23      $
 24        SUBCOM   450
 25        SUBSEQ=  .5, 1.66
 26      $
 27        SUBCOM  500
 28        SUBSEQ=  .4, -2.
 29      $
 30      BEGIN BULK

         INPUT BULK DATA CARD COUNT =      39
```

Exhibit 5.2   Case Control Deck

```
                    S O R T E D   B U L K   D A T A   E C H O
CARD
COUNT  .  1  ..  2  ..  3  ..  4  ..  5  ..  6  ..  7  ..  8
  1-  CROD   100    666    1      6
  2-  CROD   200    777    6      5
  3-  CROD   300    777    2      5
  4-  CROD   400    777    6      2
  5-  CROD   500    777    4      5
  6-  CROD   600    777    1      2
  7-  CROD   700    777    2      3
  8-  CROD   800    777    4      3
  9-  CROD   900    777    5      3
 10-  FORCE  100    6             333.   -1.    .0     .0
 11-  FORCE  200    3             70.    .0     1.     .0
 12-  FORCE  200    6             200.   .0     1.     0.
 13-  GRID   1             0.0    .0     0.            123456
 14-  GRID   2             30.0   .0     .0            3456
 15-  GRID   3             60.0   0.0    .0            3456
 16-  GRID   4             90.00  .0     .0            23456
 17-  GRID   5             60.00  30.0   .0            3456
 18-  GRID   6             30.0   30.0   .0            3456
 19-  MAT1   999999  30.0+6
 20-  PROD   666     999999  2.0
 21-  PROD   777     999999  1.5
      ENDDATA

      TOTAL COUNT=    22
```

Exhibit 5.3   Bulk Data Deck sorted by alphabetical order

```
*** USER INFORMATION MESSAGE 5293 FOR DATA BLOCK KLL

                                              EPSILONS LARGER THAN 0.001 ARE FLAGGED WITH ASTERISKS

LOAD SEQ. NO.        EPSILON        EXTERNAL WORK
      1          -6.5039855E-17     7.7121966E-02
      2           4.1069130E-17     5.8420971E-02
```

Exhibit 5.4   Epsilons

```
SIMPLE 2-D TRUSS; STATIC ANALYSIS                          JUNE  7, 1988  MSC/NASTRAN 9/ 2/87   PAGE   4
WE INTRODUCE NOW SEVERAL COMBINATIONS OF CASES

  HORIZONTAL  LOADS                                                                          SUBCASE   1

                                       D I S P L A C E M E N T   V E C T O R
POINT ID.  TYPE         T1             T2            T3            R1            R2            R3
    2       G     -1.480000E-04   2.322173E-04      0.0           0.0           0.0           0.0
    4       G     -2.960000E-04      0.0            0.0           0.0           0.0           0.0
```

Exhibit 5.5   Displacement vector for Subcase 1

```
HORIZONTAL  LOADS                                                                    SUBCASE   1

                            L O A D   V E C T O R

POINT ID.  TYPE      T1              T2        T3       R1       R2       R3
   6        G    -3.330000E+02      0.0       0.0      0.0      0.0      0.0

VERTICAL  LOADS                                                                      SUBCASE   2
POINT ID.  TYPE      T1              T2        T3       R1       R2       R3
   3        G        0.0        7.000000E+01   0.0      0.0      0.0      0.0
   6        G        0.0        2.000000E+02   0.0      0.0                SUBCOM  400

POINT ID.  TYPE      T1              T2        T3       R1       R2       R3
   3        G        0.0        7.000000E+01   0.0      0.0      0.0      0.0
   6        G    -3.330000E+02   2.000000E+02   0.0      0.0                SUBCOM  450

POINT ID.  TYPE      T1              T2        T3       R1       R2       R3
   3        G        0.0        1.162000E+02   0.0      0.0      0.0      0.0
   6        G    -1.665000E+02   3.320000E+02   0.0      0.0                SUBCOM  500

POINT ID.  TYPE      T1              T2        T3       R1       R2       R3
   3        G        0.0       -1.400000E+02   0.0      0.0      0.0      0.0
   6        G    -1.332000E+02  -4.000000E+02   0.0      0.0      0.0      0.0
```

Exhibit 5.6 Load vectors

F O R C E S   I N   R O D   E L E M E N T S   (C R O D)   SUBCASE 2

VERTICAL LOADS

ELEMENT ID.	AXIAL FORCE	TORQUE	ELEMENT ID.	AXIAL FORCE	TORQUE
400	4.333334E+01	0.0	500	1.602775E+02	0.0

Exhibit 5.7 Forces in ROD elements for Subcase 2

S T R E S S E S   I N   R O D   E L E M E N T S   (C R O D)   SUBCASE 1

HORIZONTAL LOADS

ELEMENT ID.	AXIAL STRESS	SAFETY MARGIN	TORSIONAL STRESS	SAFETY MARGIN	ELEMENT ID.	AXIAL STRESS	SAFETY MARGIN	TORSIONAL STRESS	SAFETY MARGIN
100	-7.848885E+01		0.0		200	1.480000E+02		0.0	
300	-1.046518E+02		0.0		400	7.400000E+01		0.0	
500	1.046518E+02		0.0		600	-1.480000E+02		0.0	
700	-7.400000E+01		0.0		800	-7.400000E+01		0.0	
900	0.0		0.0						

SUBCASE 2

VERTICAL LOADS

ELEMENT ID.	AXIAL STRESS	SAFETY MARGIN	TORSIONAL STRESS	SAFETY MARGIN	ELEMENT ID.	AXIAL STRESS	SAFETY MARGIN	TORSIONAL STRESS	SAFETY MARGIN
100	1.107801E+02		0.0		200	1.044444E+02		0.0	
300	-4.085506E+01		0.0		400	2.888889E+01		0.0	
500	1.068517E+02		0.0		600	-1.044444E+02		0.0	
700	-7.555556E+01		0.0		800	-7.555556E+01		0.0	
900	-4.666667E+01		0.0						

Exhibit 5.8 Stresses in ROD elements

S T R E S S E S   I N   R O D   E L E M E N T S   (C R O D)

ELEMENT ID.	AXIAL STRESS	SAFETY MARGIN	TORSIONAL STRESS	SAFETY MARGIN	ELEMENT ID.	AXIAL STRESS	SAFETY MARGIN	TORSIONAL STRESS	SAFETY MARGIN
								SUBCOM	400
100	3.229121E+01		0.0		200	2.524444E+02		0.0	
300	-1.455069E+02		0.0		400	1.028889E+02		0.0	
500	2.115035E+02		0.0		600	-2.524444E+02		0.0	
700	-1.495556E+02		0.0		800	-1.495556E+02		0.0	
900	-4.666667E+01		0.0						
								SUBCOM	450
100	1.446505E+02		0.0		200	2.473778E+02		0.0	
300	-1.201453E+02		0.0		400	8.495555E+01		0.0	
500	2.296997E+02		0.0		600	-2.473778E+02		0.0	
700	-1.624222E+02		0.0		800	-1.624222E+02		0.0	
900	-7.746667E+01		0.0						
								SUBCOM	500
100	-2.529557E+02		0.0		200	-1.496889E+02		0.0	
300	3.984940E+01		0.0		400	-2.817778E+01		0.0	
500	-1.718427E+02		0.0		600	1.496889E+02		0.0	
700	1.215111E+02		0.0		800	1.215111E+02		0.0	
900	9.333334E+01		0.0						

Exhibit 5.8 (continued)   Stresses in ROD elements

6
Problem 6

6.1 Statement of the problem

Consider once again the truss described in Figure 1.1. This example emphasizes the importance of examining the results carefully.

Assume that the truss is subjected to the following loading conditions: first, at GRID 2, $F_x = 600$ lb; at GRID 3, $F_x = 200$ lb; at GRID 4, $F_x = 60$ lb; and at GRID 6, $F_x = F_y = 125$ lb; and second, at GRID 3, $F_x = 600$ lb; at GRID 5, $F_y = -150,000$ lb; and at GRID 6, $F_x = F_y = 375$ lb.

Furthermore, assume that the material (steel) has a yield stress (σ_y) equal to 30,000 psi. This means that if the normal stress in the material reaches 30,000 psi, the structure deforms permanently and can no longer be considered linear. Determine the stresses in all RODs.

6.2 Cards introduced

Case Control Deck   None

Bulk Data Deck   LOAD
                              Continuation Cards

6.3 MSC/NASTRAN formulation

The model (shown in Figure 1.2) has been described in the previous examples. The Bulk Data Deck features a new card, the LOAD card, which is used to combine several types of loadings. In addition, the material yield stress is included on the MAT1 card. This is used to compute safety margin coefficients.

6.4 Input Data Deck

Exhibit 6.1 shows the Executive Control Deck. The Case Control Deck (Exhibit 6.2) specifies two subcases. No plots are created. The stresses are printed for all members in both subcases.

The individual forces acting on the truss are defined by the five FORCE cards included in the Bulk Data Deck (Exhibit 6.3). The first LOAD card of the Bulk Data Deck (9999) is selected for the first subcase by the LOAD = 9999 card in the Case Control Deck. This card combines the FORCE cards necessary to describe the first loading configuration. The second LOAD card, selected by LOAD = 555555 in the Case Control Deck, combines the FORCE cards that describe the second loading condition. Note that the LOAD card allows the use of different scale factors. The first, which goes in field 3, affects all the FORCE cards referenced on the LOAD card; each remaining factor, which goes in fields 4, 6, etc., affects only the FORCE card specified in the subsequent field.

The MAT1 card includes stress limit information for tension and compression. A yield stress value of 30,000 psi is specified. MSC/NASTRAN then calculates the safety margin for the stresses in each element.

Some Bulk Data Deck cards may require two or more "physical" cards. In this case, the continuation card feature must be used. It works as follows: in the last field (the tenth field) of the first card we type a + followed by an arbitrary "word". Then, in the first field of the continuation card we type the same thing (a + followed by the same "word"). This indicates that these two cards go together as one entity. The MAT1 card demonstrates this feature.

6.5 Results

The output file shows two fairly small EPSILONs (Exhibit 6.5). This is a necessary condition (but not sufficient, as will be seen) to consider the results valid.

Exhibit 6.6 depicts the values of the stresses for the RODs in both subcases. In the first subcase all the stresses are below the yield value (30,000 psi) indicating that the truss behaved linearly. The safety margin, defined as

$$S.M. = (\sigma_y/\sigma_a) - 1.0 \qquad (6.1)$$

where σ_a represents the actual stress in the member, is positive for all the RODs. This indicates that the maximum admissible value (yield stress) has not yet been reached.

Now examine the stresses for subcase 2. For all RODs, the stresses are greater (in absolute value) than the yield stress. Consequently, the safety margins are negative, suggesting the possibility of structural failure. Strictly speaking, this indicates that the results obtained for the second subcase are not valid because they were obtained assuming linear behavior. This implies that a nonlinear analysis is required to accurately estimate the response of the structure under these loads. Notice that MSC/NASTRAN does not give any warnings regarding the validity of the linear behavior assumption. It is the user's responsibility to interpret the output correctly and verify that this assumption is not violated.

```
N A S T R A N   E X E C U T I V E   C O N T R O L   D E C K   E C H O

   ID    PROBLEM,SIX
   SOL   24
   TIME  5
   CEND
```

Exhibit 6.1   Executive Control Deck

```
          C A S E   C O N T R O L   D E C K   E C H O
   CARD
   COUNT
     1       TITLE=  SIMPLE 2-D TRUSS; STATIC ANALYSIS
     2       ECHO= BOTH
     3       $
     4       STRESS=ALL
     5       $
     6       SUBCASE  1
     7       LOAD =9999
     8       $
     9       SUBCASE  2
    10       LOAD= 555555
    11       $
    12       BEGIN BULK
```

Exhibit 6.2   Case Control Deck

```
                    I N P U T   B U L K   D A T A   D E C K   E C H O
  .   1  ..   2  ..   3  ..   4  ..   5  ..   6  ..   7  ..   8  ..   9
  $
  $ HERE WE COMBINE SEVERAL  LOADS
  $
  LOAD    9999    1.00    1.00    22      1.5     23      6.0     24
  LOAD    555555  2.00    1.5     22      3.0     25
  $
  FORCE   22      6               250.0   .5      .5      .0
  FORCE   22      3               200.    1.      .0      .0
  FORCE   23      4               40.00   1.0     .0      .0
  FORCE   24      2               100.0   1.0     .0      .0
  FORCE   25      5               25.+3   .0      -1.     .0
  $
  GRID,1,  ,0.0, .0, 0., ,123456
  GRID,2,  ,30.0,  .0, .0,,3456
  GRID,3,, 60.0,  0.0, .0,,3456
  GRID,4,  ,90.00, .0, .0,  ,23456
  GRID,5,  ,60.00, 30.0,  .0,,3456
  GRID,6,,30.0, 30.0,  .0,, 3456
  $
  $ NOW WE SPECIFY THE CONNECTIVITY OF THE ROD ELEMENTS
  $
  CROD    100     666     1       6
  CROD    200     777     6       5
  CROD    300     777     2       5
  CROD    400     777     6       2
  CROD    900     777     5       3
  CROD    500     777     4       5
  CROD    600     777     1       2
  CROD    700     777     2       3
  CROD    800     777     4       3
  $
  $ GEOMETRIC PROPERTIES OF THE RODS
  $
  PROD,666,999999,2.0
  PROD    777     999999  1.5
```

Exhibit 6.3   Input Bulk Data Deck

```
$ MATERIAL PROPERTIES=== WE PUT SIGMA MAX ON THE CONTINUATION  CARD
$                        TO DETERMINE SAFETY FACTORS
$
MAT1    999999  30.0+6                                              +PLPL
+PLPL   30.+3   30.00+3
$
ENDDATA
```

Exhibit 6.3 (continued) Input Bulk Data Deck

```
                          S O R T E D   B U L K   D A T A   E C H O
CARD
COUNT   .   1 ..  2 .. 3 .. 4 .. 5 .. 6 .. 7 .. 8 .. 9 .. 10 .
1-      CROD  100  666  1   6
2-      CROD  200  777  6   5
3-      CROD  300  777  2   5
4-      CROD  400  777  6   2
5-      CROD  500  777  4   5
6-      CROD  600  777  1   2
7-      CROD  700  777  2   3
8-      CROD  800  777  4   3
9-      CROD  900  777  5   3
```

Exhibit 6.4 Bulk Data Deck sorted by alphabetical order

Seq	Card	1	2	3	4	5	6	7	8	Cont.
10-	FORCE	22	3		200.	1.	.0	.0		
11-	FORCE	22	6		250.0	.5	.5	.0		
12-	FORCE	23	4		40.00	1.0	.0	.0		
13-	FORCE	24	2		100.0	1.0	.0	.0		
14-	FORCE	25	5		25.+3	.0	-1.	.0		
15-	GRID	1		0.0	.0	0.		123456		
16-	GRID	2		30.0	.0	.0		3456		
17-	GRID	3		60.0	0.0	.0		3456		
18-	GRID	4		90.00	.0	.0		23456		
19-	GRID	5		60.00	30.0	.0		3456		
20-	GRID	6		30.0	30.0	.0		3456		
21-	LOAD	9999	1.00	1.00	22	1.5	23	6.0	24	
22-	LOAD	555555	2.00	1.5	22	3.0	25			
23-	MAT1	999999	30.0+6							+PLPL
24-	+PLPL	30.+3	30.00+3							
25-	PROD	666	999999	2.0						
26-	PROD	777	999999	1.5						
	ENDDATA									

Exhibit 6.4 (continued) Bulk Data Deck sorted by alphabetical order

** USER INFORMATION MESSAGE 5293 FOR DATA BLOCK KLL

EPSILONS LARGER THAN 0.001 ARE FLAGGED WITH ASTERISKS

LOAD SEQ. NO.	EPSILON	EXTERNAL WORK
1	4.2531336E-18	2.8131521E-01
2	-3.7226701E-17	2.2753369E+04

Exhibit 6.5 Epsilons

S T R E S S E S   I N   R O D   E L E M E N T S   (C R O D)

SUBCASE   1

ELEMENT ID.	AXIAL STRESS	SAFETY MARGIN	TORSIONAL STRESS	SAFETY MARGIN		ELEMENT ID.	AXIAL STRESS	SAFETY MARGIN	TORSIONAL STRESS	SAFETY MARGIN
100	8.838835E+01	3.4E+02	0.0			200	1.421085E-14	2.1E+18	0.0	
300	-7.105427E-15	4.2E+18	0.0			400	0.0		0.0	
500	1.421085E-14	2.1E+18	0.0			600	5.733333E+02	5.1E+01	0.0	
700	1.733333E+02	1.7E+02	0.0			800	4.000000E+01	7.5E+02	0.0	
900	0.0									

S T R E S S E S   I N   R O D   E L E M E N T S   (C R O D)

SUBCASE   2

ELEMENT ID.	AXIAL STRESS	SAFETY MARGIN	TORSIONAL STRESS	SAFETY MARGIN		ELEMENT ID.	AXIAL STRESS	SAFETY MARGIN	TORSIONAL STRESS	SAFETY MARGIN
100	-3.509018E+04	-1.5E-01	0.0			200	-3.333333E+04	-1.0E-01	0.0	
300	-4.714045E+04	-3.6E-01	0.0			400	3.333333E+04	-1.0E-01	0.0	
500	-9.428091E+04	-6.8E-01	0.0			600	3.373333E+04	-1.1E-01	0.0	
700	6.706666E+04	-5.5E-01	0.0			800	6.666666E+04	-5.5E-01	0.0	
900	0.0									

Exhibit 6.6   Stresses in ROD elements

7
Problem 7

7.1 Statement of the problem

Consider the cantilever beam shown in Figure 7.1. It has a constant cross section and it is subjected to three load conditions:

1   Its own weight (gravity load),

2   A force of 45 lb in the negative y-direction and a moment of 233 in-lb in the z-direction, both acting at the beam's free end, and

3   A distributed load applied to a portion of the beam as shown Figure 7.1.

Assume E = 30,000,000.00 psi, ν = 0.333, and ρ (mass per unit of volume) = 7×10^{-4} lb-s^2/in^4. The value of K (shear factor) for this cross section is 5/6. Compute the stresses and deflections for the three load conditions.

7.2 Cards introduced

Case Control Deck   None

Bulk Data Deck   CBAR
                             GRAV
                             MOMENT
                             PBAR
                             PLOAD1

7.3 MSC/NASTRAN formulation

The structure is modeled with BAR elements. The BAR is a beam element with constant cross section, and coincident neutral and shear axes.

(MSC/NASTRAN includes another beam element, called BEAM, which is a more sophisticated version of the BAR element.)

Four BAR elements are used to model this structure. In addition, three subcases are defined, one for each loading condition. The model is shown in Figure 7.2.

7.4 Input Data Deck

Notice that SOL 24 is selected in the Executive Control Deck (Exhibit 7.1). This is still a linear static analysis of the form

$$[K]\{x\} = \{f\}. \tag{7.1}$$

The fact that a different finite element is used (BAR instead of ROD) does not change this.

The Case Control Deck (Exhibit 7.2) specifies three subcases. The displacement vector at the free end (GRID 5) and the stresses for BAR element 10 (the element closest to the support) are printed for all subcases. This is due to the definition of SET 100 and SET 500.

The plot deck is similar to those of previous examples. The axes have been chosen (AXES Z, X, Y) to project the structure onto the x-y plane. Exhibits 7.3 and 7.4 show the Bulk Data Deck.

Examine the GRID cards. At GRID 1, all displacements and rotations are constrained (123456 in field 8), because the beam is clamped at this GRID point. At the other GRID points only the translation in the y-direction and the rotation about z are left free (1345 in field 8). Since no loads act in the x-direction or z-direction, we know beforehand that the beam will only exhibit displacements in the y-direction and rotations about the z-axis.

The connectivity of the BAR elements is defined with CBAR cards. The second field of the CBAR card is the ID number, which must be unique for each element. The third field (3333) points to a PBAR card. Fields 4 and 5 define GA and GB, the GRIDs that specify the initial and final node of the BAR. Note that the sequence in which GA and GB are specified does not affect the numerical results. It only affects the way in which the results are printed and interpreted. Fields 6, 7, and 8, give the components of the V-vector. This is a very important concept that is explained in the following paragraph.

Two pieces of information are required to define a BAR element: the end points of the BAR, and the orientation of the principal axes. The end points of the BAR are defined by GA and GB on the CBAR card. The orientation of the principal axes is defined by specifying two planes: Plane 1 and Plane 2.

Consider Plane 1 (remember that a plane can be defined by two vectors). Assume that one vector is the vector extending from one end of the beam to the other --the (GA, GB) vector in this case. The second vector will be the V-vector, a vector whose origin is at one end of the beam (A), and has components in the x-, y-, and z-direction (measured from end A and parallel to the displacement coordinate system) given by the entries X_1, X_2, and X_3 on the CBAR card (fields 6, 7, and 8). These two vectors, (GA, GB) and the V-vector, uniquely determine Plane 1 (assuming that (GA, GB) and V do not coincide).

Plane 2 is automatically defined as the plane perpendicular to Plane 1, that passes through the end points of the BAR element. This situation is explained more graphically in Figure 7.3. Note that when defining Plane 1 and Plane 2, an element coordinate system associated with that particular BAR element is implicitly defined. This element coordinate system (y_e, z_e) is employed to specify the locations on the cross section of the beam where the stresses are to be computed. It is also used to define the element shear and bending moments for output purposes.

Finally, I_1 designates the moment of inertia of the cross section of the beam *with respect to an axis perpendicular to Plane 1*. I_2 denotes the moment of inertia with respect to an axis perpendicular to Plane 2.

In this example the x-z plane is to be defined as Plane 1. This is accomplished by defining V as (1.0, 0.0, 1.0). Many other choices for V would have achieved the same result, for instance, (444.0, 0.0, 777.), (345678., 0.0, 65444.8), etc. They all define, in combination with (GA, GB), the same plane.

The PBAR card is used to supply geometric information (cross section properties) regarding the BAR elements. Since all the beam elements in this example share the same properties, only one PBAR card is necessary. The second field (3333) of the PBAR card is the ID card and the third field points to the MAT1 card. The first continuation card is used to give the locations (coordinates in the element coordinate system) of the points at which the stresses are to be computed; in this case, these correspond to the corner points on the cross section of the beam. This information is not required if there is not stress output request in the Case Control Deck.

The coefficients K_1 and K_2 are needed only if shear stiffness is to be included in the computation of the deflection of the beam. Shear stiffness is normally negligible in slender Euler-Bernoulli beams since the deflections caused by bending are several orders of magnitude larger than the deflections caused by shear. In this example, however, these coefficients are included to demonstrate this feature of the PBAR card. Remember that the shear displacement of a beam (δ_s) at a location ξ is given by

$$\delta_s = [(Q(\xi)l)/(KAG)] \qquad (7.2)$$

where $Q(\xi)$ is the shear force at ξ, l is the length of the beam, A is the area of the cross section, G is the beam shear modulus, and K is the shear factor.

For a rectangular section K is equal to 5/6 (.833). For other sections the value of K can be obtained from a standard table on properties of beams.

The MAT1 card incorporates the value of ρ, mass per unit of volume (NOT per unit of length) which is needed to define the gravity load. This value goes in the sixth field (.0007 lb-s^2/in^4).

Consider the load cards. The GRAV card specifies that the gravity is acting in the negative y-direction and its value is 386 in/s^2. This card is selected in the first subcase by LOAD = 1 in the Case Control Deck.

LOAD = 2 selects two cards for the second subcase. The FORCE card has been already explained. The format of the MOMENT card is analogous to that of the FORCE card except that it defines a moment instead of a force.

The PLOAD1 card selected for the third subcase is also used to define loads acting on beam elements. The third field entry (40) indicates that the load is acting on BAR element number 40. The FY in the fourth field indicates that this is a force (F) acting in the y-direction (Y). The FR in the next field means that the "fractional" option is chosen to define the initial and final locations of the distributed load on the BAR element. X_1 = 0.5 and P_1 = -10.0 specify that at a location halfway between GRID 4 and GRID 5 the intensity of the load is -10 lb/in. X_2 = 1.0 and P_2 = -100 specify that at GRID 5 the intensity of the load is -100 lb/in. X_1 = .5 results from (262.5-225)/75 = 0.5, and X_2 = 1.0 results from (300-225)/75 = 1.0 (262.5 represents the coordinate of a point half way between GRID 4 and GRID 5, 75 is the length of each BAR element).

7.5 Results

The EPSILONs are shown in Exhibit 7.5 They are small, indicating that the problem is numerically stable. Exhibit 7.6 shows the displacement at the free end of the beam (GRID 5) for the three subcases. Exhibit 7.7 shows the value of the stresses for BAR element 10, for each subcase. These stresses are calculated at the points on the cross section of the beam specified on the PBAR card. SXC, SXD, SXE, and SXF are stresses due to bending only. Under the AXIAL heading there are only zeroes. This is because this column corresponds to axial stresses produced by tension only. Since there are no forces acting in the x-direction, no stresses of this type are produced. S-MAX and S-MIN show the axial stresses due to the combined effect of bending and tension (maximum and minimum values). STATION 0.000 refers to GA, the initial node of the BAR, and STATION 1.000 refers to GB, the final node of the BAR.

The first plot (Figure 7.4) shows the finite element model. The other three plots (Figures 7.5, 7.6, and 7.7), generated with only one instruction (statement 29 of the Case Control Deck) show the deformed and the undeformed structure for each subcase.

7.6 Additional comments

In this example, I_1 (the entry equal to 104.20 on the PBAR card) could have been omitted since the displacement in the z-direction and the rotation about y were constrained. If these degrees of freedom were released on the GRID cards (degrees of freedom 3 and 5) the same results would have been obtained since there were no loads acting in the z-direction, and the beam had stiffness in that direction due to the fact that I_1 was not zero.

Note also that no torsional moment was applied to the beam, and for this reason no rotations about x were expected. Consequently, no value for J was defined on the PBAR card, and degree of freedom 4 (rotation about x) was restrained on the GRID cards. If a non-zero value of J had been specified, degree of freedom 4 could have been released obtaining the same results, since the beam would have had torsional stiffness in that direction while no torsional moment was applied.

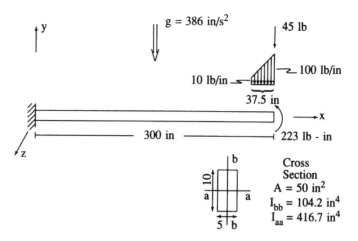

Figure 7.1

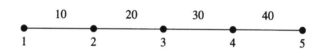

Figure 7.2

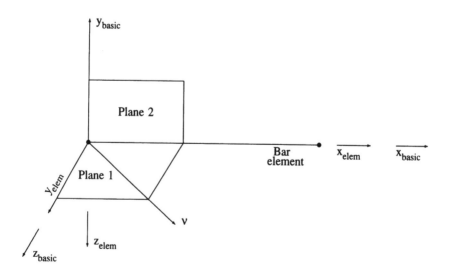

Figure 7.3

```
N A S T R A N   E X E C U T I V E   C O N T R O L   D E C K   E C H O
    ID    BEAM,ONE
    SOL   24
    TIME  5
    CEND
```

Exhibit 7.1   Executive Control Deck

```
                C A S E    C O N T R O L   D E C K   E C H O
    CARD
     1      TITLE= ANALYSIS OF BEAM
     2      $
     3      SET  500=  5
     4      SET 100=   10
     5      STRESS=    100
     6      DISPLACEMENT=   500
     7      $
     8      ECHO=  BOTH
     9      $
    10      SUBCASE  1
    11      LABEL= GRAVITY LOAD
    12      LOAD =1
    13      $
    14      SUBCASE  2
    15      LABEL= MOMENT AND FORCE AT TIP
    16      LOAD  =2
    17      $
    18      SUBCASE  3
    19      LABEL= DISTRIBUTED LOAD
    20      LOAD =3
    21      $
    22      $$$$$$$$$$$$$$$$$   P L O T S $$$$$$$$$$$$$$$$$$$$$$
    23      OUTPUT(PLOT)
    24      SET  200  INCLUDE  ALL
    25      AXES  Z,  X,  Y
    26      VIEW .0,.0,.0
    27      FIND
    28      PLOT  LABEL  BOTH
    29      PLOT  STATIC  DEFORMATION  0   SET   200
    30      BEGIN BULK
```

Exhibit 7.2   Case Control Deck

```
                  I N P U T   B U L K   D A T A   D E C K   E C H O
   .  1 ..  2 ..  3 ..  4 ..  5 ..  6 ..  7 ..  8 ..  9 .. 10 .
   $
   $ GRID POINTS
   $
   GRID    1              .0      .0      .0                 123456
   GRID    2           75.00      .0      .0                 1345
   GRID    3           150.0      .0      .0                 1345
   GRID    4            225.      .0      .0                 1345
   GRID    5            300.      .0      .0                 1345
   $
   $ WE USE CBAR CARDS TO SPECIFY CONNECTIVITY OF BEAMS
   $
   CBAR   10    3333     1       2       1.      .0      1.
   CBAR   20    3333     2       3       1.      0.      1.
   CBAR   30    3333     3       4       1.      .0      1.
   CBAR   40    3333     4       5       1.0     .00     1.
   $
   $ PBAR ( WE WILL INCLUDE SHEAR DEFORMATIONS K=.833 = 5/6 )
   $
   PBAR 3333   1212    50.00   104.20  416.70                      +LLLL
   +LLLL 2.500  5.00   -2.50    5.00    2.5    -5.    -2.5    -5.   +LK
   +LK   .833   .833
```

Exhibit 7.3   Input Bulk Data Deck

```
$
$   NOW IN THE MAT1 CARD WE HAVE TO PUT ALSO RHO (MASS/VOL.)
$   SINCE WE WILL APPLY A GRAVITY LOAD
$
MAT1    1212    30.+6           .333    7.-4
$
$$$$$$  WE APPLY THE  LOADS  $$$$$$$$$$$$$$
GRAV    1               -386.   .00     1.00    .0
FORCE   2       5               45.     .0      -1.0    .0
MOMENT  2       5               233.    .0      .0      1.
PLOAD1  3       40      FY      FR      .5      -10.0   1.0     -100.
$
ENDDATA
```

Exhibit 7.3 (continued) Input Bulk Data Deck

S O R T E D   B U L K   D A T A   E C H O

CARD COUNT	.	1	2	3	4	5	6	7	8	9	10
1-	CBAR	10	3333	1	2	1.	.0	1.			
2-	CBAR	20	3333	2	3	1.	0.	1.			
3-	CBAR	30	3333	3	4	1.	.0	1.			
4-	CBAR	40	3333	4	5	1.0	.00	1.			
5-	FORCE	2	5		-386.	45.	.0	-1.0	.0		
6-	GRAV	1		.0	.00	1.00	.0				
7-	GRID	1		.0	.0	.0	.0	123456			
8-	GRID	2		75.00	.0	.0		1345			
9-	GRID	3		150.0	.0	.0		1345			
10-	GRID	4		225.	.0	.0		1345			
11-	GRID	5		300.	.0	.0		1345			
12-	MAT1	1212	30.+6		.333	7.-4					
13-	MOMENT	2	5		233.	.0		1.			
14-	PBAR	3333	1212	50.00	104.20	416.70	.0	1.			+LLLL
15-	+LLLL	2.500	5.00	-2.50	5.00	2.5	-5.	-2.5	-5.		+LK
16-	+LK	.833	.833					1.0			
17-	PLOAD1	3	40	FY	FR	.5	-10.0	-100.			
	ENDDATA										

Exhibit 7.4   Bulk Data Deck sorted by alphabetical order

```
** USER INFORMATION MESSAGE 5293 FOR DATA BLOCK KLL

LOAD SEQ. NO.      EPSILON           EXTERNAL WORK      EPSILONS LARGER THAN 0.001 ARE FLAGGED WITH ASTERISKS
      1         -1.1099577E-15       9.1988452E+02
      2          3.4372411E-15       6.9249821E-01
      3          1.4223149E-15       1.3330730E+03
```

Exhibit 7.5 Epsilons

```
GRAVITY LOAD                                                    SUBCASE   1

                          D I S P L A C E M E N T   V E C T O R
POINT ID.  TYPE    T1         T2          T3          R1          R2          R3
      5     G     0.0     -1.118316E+00   0.0         0.0         0.0     -5.015186E-03

MOMENT AND FORCE AT TIP                                         SUBCASE   2

                          D I S P L A C E M E N T   V E C T O R
POINT ID.  TYPE    T1         T2          T3          R1          R2          R3
      5     G     0.0     -3.158748E-02   0.0         0.0         0.0     -1.563955E-04

DISTRIBUTED LOAD                                               SUBCASE   3

                          D I S P L A C E M E N T   V E C T O R
POINT ID.  TYPE    T1         T2          T3          R1          R2          R3
      5     G     0.0     -1.385084E+00   0.0         0.0         0.0     -6.772310E-03
```

Exhibit 7.6 Displacement vector at GRID 5

GRAVITY LOAD   SUBCASE 1

S T R E S S   D I S T R I B U T I O N   I N   B A R   E L E M E N T S   (C B A R)

ELEMENT ID.	STATION (PCT)	SXC	SXD	SXE	SXF	AXIAL	S-MAX	S-MIN
10	0.000	-7.294816E+03	-7.294816E+03	7.294816E+03	7.294816E+03	0.0	7.294816E+03	-7.294816E+03
10	1.000	-4.103334E+03	-4.103334E+03	4.103334E+03	4.103334E+03	0.0	4.103334E+03	-4.103334E+03

MOMENT AND FORCE AT TIP   SUBCASE 2

S T R E S S   D I S T R I B U T I O N   I N   B A R   E L E M E N T S   (C B A R)

ELEMENT ID.	STATION (PCT)	SXC	SXD	SXE	SXF	AXIAL	S-MAX	S-MIN
10	0.000	-1.591913E+02	-1.591913E+02	1.591913E+02	1.591913E+02	0.0	1.591913E+02	-1.591913E+02
10	1.000	-1.186945E+02	-1.186945E+02	1.186945E+02	1.186945E+02	0.0	1.186945E+02	-1.186945E+02

DISTRIBUTED LOAD   SUBCASE 3

S T R E S S   D I S T R I B U T I O N   I N   B A R   E L E M E N T S   (C B A R)

ELEMENT ID.	STATION (PCT)	SXC	SXD	SXE	SXF	AXIAL	S-MAX	S-MIN
10	0.000	-7.086933E+03	-7.086933E+03	7.086933E+03	7.086933E+03	0.0	7.086933E+03	-7.086933E+03
10	1.000	-5.230832E+03	-5.230832E+03	5.230832E+03	5.230832E+03	0.0	5.230832E+03	-5.230832E+03

Exhibit 7.7 Stresses in BAR element 10

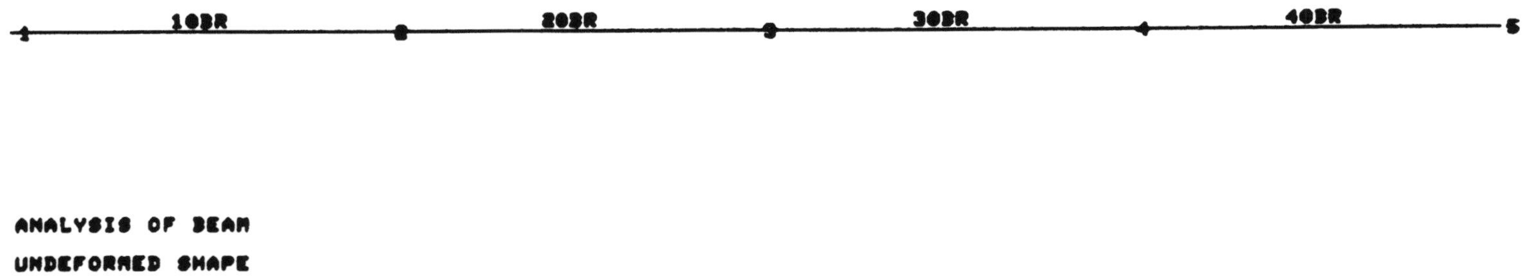
1
10BR
20BR
30BR
40BR
5
ANALYSIS OF BEAM
UNDEFORMED SHAPE

Figure 7.4 Finite element model of the beam using BAR elements.

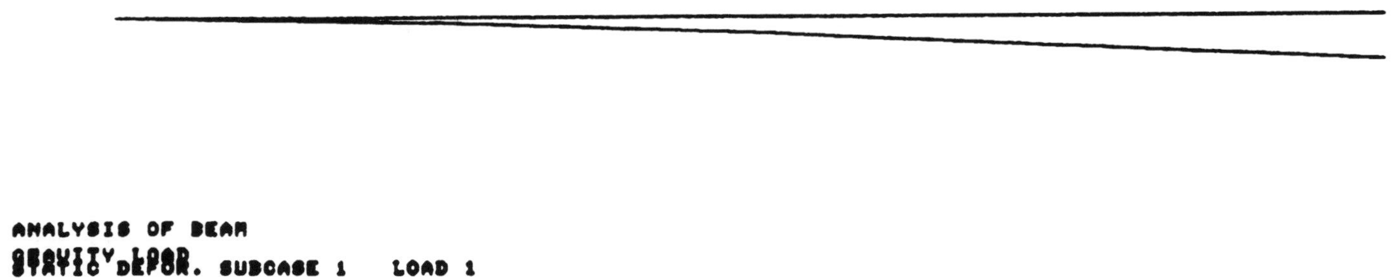
ANALYSIS OF BEAM
GRAVITY LOAD
STATIC DEFOR. SUBCASE 1 LOAD 1

Figure 7.5 Deformed shape of the beam (gravity load).

ANALYSIS OF BEAM
LOAD 2

Figure 7.6 Deformed shape of the beam (moment of force at the tip).

ANALYSIS OF BEAM
LOAD 3

Figure 7.7 Deformed shape of the beam (distributed load).

8
Problem 8

8.1 Statement of the problem

Analyze the same beam structure of Problem 7 (Figures 7.1 and 7.2) but neglect the deformation due to shear when computing the displacement vector.

8.2 Cards introduced

Case Control Deck   None

Bulk Data Deck   None

8.3 MSC/NASTRAN formulation

The same model and data deck of Problem 7 are used. The only modification is that the shear factor K is not included on the PBAR card.

8.4 Input Data Deck

Exhibit 8.1 shows the Executive Control Deck and Exhibit 8.2 shows the Case Control Deck. The plot statements in the Case Control Deck have been deleted.

Exhibits 8.3 and 8.4 show the Bulk Data Deck. Notice in Exhibit 8.3 the only modification made to the Bulk Data Deck used for Problem 7: the fields corresponding to K_1 and K_2, the shear factors, are left blank on the PBAR card. Thus, the effect of shear is neglected in the calculation of the beam displacement.

8.5 Results

Exhibit 8.5 shows the three values of EPSILON, which are small but different than the values obtained in Problem 7 (Exhibit 7.5).

Notice that the values obtained for the displacement in the y-direction (Exhibit 8.6) are slightly different compared to those obtained in Problem 7. Problem 7 shows larger values (Exhibit 7.6) because the displacement vector in that case included the deformations due to shear. Rotations about z (R3), however, are the same in both cases since they are not affected by K_1 or K_2.

The values printed for the stresses (Exhibit 8.7) are identical to those calculated in Problem 7 (Exhibit 7.7) since these stresses are the result of bending and tension only. They do not depend on the factor K, which only affects the transverse translational components of the displacement vector.

```
N A S T R A N   E X E C U T I V E   C O N T R O L   D E C K   E C H O
     ID    BEAM,ONE
     SOL   24
     TIME  5
     CEND
```

Exhibit 8.1   Executive Control Deck

```
                  C A S E    C O N T R O L    D E C K    E C H O
CARD
COUNT
  1       TITLE= ANALYSIS OF BEAM
  2       $
  3       SET  500=  5
  4       SET 100=   10
  5       STRESS=    100
  6       DISPLACEMENT=    500
  7       $
  8       ECHO=  BOTH
  9       $
 10       SUBCASE  1
 11       LABEL= GRAVITY LOAD
 12       LOAD =1
 13       $
 14       SUBCASE  2
 15       LABEL= MOMENT AND FORCE AT TIP
 16       LOAD  =2
 17       $
 18       SUBCASE  3
 19       LABEL= DISTRIBUTED LOAD
 20       LOAD =3
 21       $
 22       BEGIN BULK
```

Exhibit 8.2   Case Control Deck

```
              I N P U T   B U L K   D A T A   D E C K   E C H O
  .   1  ..   2  ..   3  ..   4  ..   5  ..   6  ..   7  ..   8  ..   9  ..  10  .
$ GRID POINTS
GRID     1                .0        .0        .0                  123456
GRID     2             75.00        .0        .0                  1345
GRID     3             150.0        .0        .0                  1345
GRID     4              225.        .0        .0                  1345
GRID     5              300.        .0        .0                  1345
$ WE USE CBAR CARDS TO SPECIFY CONNECTIVITY OF BEAMS
CBAR    10  3333     1         2         1.        .0        .0        1.
CBAR    20  3333     2         3         1.        0.                  1.
CBAR    30  3333     3         4         1.        .0                  1.
CBAR    40  3333     4         5         1.0       .00                 1.
$ PBAR ( WE DO NOT INCLUDE SHEAR DEFORMATIONS )
PBAR  3333  1212  50.00    104.20    416.70                            +LLLL
+LLLL 2.500  5.00  -2.50     5.00      2.5      -5.       -2.5      -5.
$ NOW IN THE MAT1 CARD WE HAVE TO PUT ALSO RHO (MASS/VOL.)
$ SINCE WE WILL APPLY A GRAVITY LOAD
MAT1  1212  30.+6           .333      7.-4
$$$$$$ WE APPLY THE LOADS $$$$$$$$$$$$$$
GRAV     1         -386.    .00       1.00      .0
FORCE    2     5     45.    .0                  -1.0       .0
MOMENT   2     5    233.    .0                  .0         1.
PLOAD1   3    40    FY      FR        .5        -10.0      1.0     -100.
ENDDATA
```

Exhibit 8.3 Input Bulk Data Deck

S O R T E D   B U L K   D A T A   E C H O

CARD COUNT	.	1	2	3	4	5	6	7	8	9	10	.
1-	CBAR	10	3333	1	2	1.	.0		1.			
2-	CBAR	20	3333	2	3	1.	0.		1.			
3-	CBAR	30	3333	3	4	1.	.0		1.			
4-	CBAR	40	3333	4	5	1.0	.00		1.			
5-	FORCE	2	5		45.	.0	-1.0		.0			
6-	GRAV	1		-386.	.00	1.00	.0					
7-	GRID	1		.0	.0	.0			123456			
8-	GRID	2		75.00	.0	.0			1345			
9-	GRID	3		150.0	.0	.0			1345			
10-	GRID	4		225.	.0	.0			1345			
11-	GRID	5		300.	.0	.0			1345			
12-	MAT1	1212	30.+6		.333	7.-4						
13-	MOMENT	2	5	233.	.0	.0			1.			
14-	PBAR	3333	1212	50.00	104.20	416.70					+LLLL	
15-	+LLLL	2.500	5.00	-2.50	5.00	2.5	-5.	-2.5	-5.			
16-	PLOAD1	3	40	FY	FR	.5	-10.0	1.0	-100.			
	ENDDATA											

Exhibit 8.4 Bulk Data Deck sorted by alphabetical order

```
*** USER INFORMATION MESSAGE 5293 FOR DATA BLOCK KLL

LOAD SEQ. NO.      EPSILON        EXTERNAL WORK      EPSILONS LARGER THAN 0.001 ARE FLAGGED WITH ASTERISKS
        1        7.1582304E-16    9.1815948E+02
        2        1.7492873E-15    6.9185013E-01
        3        3.2287145E-15    1.3318256E+03
```

Exhibit 8.5   Epsilons

GRAVITY LOAD   SUBCASE   1

                              D I S P L A C E M E N T   V E C T O R
POINT ID. TYPE   T1   T2   T3   R1   R2   R3
    5   G   0.0   -1.117019E+00   0.0   0.0   0.0   -5.015186E-03

MOMENT AND FORCE AT TIP   SUBCASE   2

                              D I S P L A C E M E N T   V E C T O R
POINT ID. TYPE   T1   T2   T3   R1   R2   R3
    5   G   0.0   -3.155867E-02   0.0   0.0   0.0   -1.563955E-04

DISTRIBUTED LOAD   SUBCASE   3

                              D I S P L A C E M E N T   V E C T O R
POINT ID. TYPE   T1   T2   T3   R1   R2   R3
    5   G   0.0   -1.383824E+00   0.0   0.0   0.0   -6.772310E-03

Exhibit 8.6   Displacement vector at GRID 5

GRAVITY LOAD   SUBCASE 1

STRESS DISTRIBUTION IN BAR ELEMENTS   (C B A R)

ELEMENT ID.	STATION (PCT)	SXC	SXD	SXE	SXF	AXIAL	S-MAX	S-MIN
10	0.000	-7.294816E+03	-7.294816E+03	7.294816E+03	7.294816E+03	0.0	7.294816E+03	-7.294816E+03
10	1.000	-4.103334E+03	-4.103334E+03	4.103334E+03	4.103334E+03	0.0	4.103334E+03	-4.103334E+03

MOMENT AND FORCE AT TIP   SUBCASE 2

STRESS DISTRIBUTION IN BAR ELEMENTS   (C B A R)

ELEMENT ID.	STATION (PCT)	SXC	SXD	SXE	SXF	AXIAL	S-MAX	S-MIN
10	0.000	-1.591913E+02	-1.591913E+02	1.591913E+02	1.591913E+02	0.0	1.591913E+02	-1.591913E+02
10	1.000	-1.186945E+02	-1.186945E+02	1.186945E+02	1.186945E+02	0.0	1.186945E+02	-1.186945E+02

DISTRIBUTED LOAD   SUBCASE 3

STRESS DISTRIBUTION IN BAR ELEMENTS   (C B A R)

ELEMENT ID.	STATION (PCT)	SXC	SXD	SXE	SXF	AXIAL	S-MAX	S-MIN
10	0.000	-7.086933E+03	-7.086933E+03	7.086933E+03	7.086933E+03	0.0	7.086933E+03	-7.086933E+03
10	1.000	-5.230832E+03	-5.230832E+03	5.230832E+03	5.230832E+03	0.0	5.230832E+03	-5.230832E+03

Exhibit 8.7 Stresses in BAR element 10

9
Problem 9

9.1 Statement of the problem

Analyze the two-beam structure depicted in Figure 9.1. Both beams are clamped at the ends. Figure 9.2 shows the loads acting on the structure. Consider a value of $E=30\times10^6$ psi and a value of $\nu=.333$. Determine the deformed shape of the structure, the reactions, and the forces and moments in the beams.

9.2 Cards introduced

<u>Case Control Deck</u>   None

<u>Bulk Data Deck</u>   None

9.3 MSC/NASTRAN formulation

The model, which consists of six BAR elements, is shown in Figure 9.3. Two new features of the CBAR card are demonstrated: the use of pin flags to represent hinged connections, and the use of G0 as an alternative to the V-vector to define Plane 1.[1]

---

[1] Problem 7 introduced the concept of Planes 1 and 2 to define the orientation of a BAR element.

9.4 Input Data Deck

Exhibit 9.1 shows the Executive Control Deck and Exhibit 9.2 shows the Case Control Deck.

The Case Control Deck should be familiar at this stage. The VIEW card of the plot subdeck selects $(\alpha, \ \beta, \ \gamma) = (0°, \ 25°, \ 30°)$ to get a three-dimensional view of the structure. See Figure 1.3 for the convention used to select the projection plane and the rotations. The choice of VIEW 0.0,0.0,0.0, as used in previous examples, would have not been appropriate since the I beam would then have reduced to a single point on the plot. Unfortunately, it is not always easy to select the correct values for the rotations.

Exhibits 9.3 and 9.4 show the Bulk Data Deck.

The GRID cards corresponding to the supports (GRID 1, 5, 50, and 51) show all six degrees of freedom constrained (123456 in field 8 of the GRID card), consistent with the fact that the beams are clamped at these points. At GRID 2, 3, and 4 the beam is free to move in any direction (rotations and translations); thus, no constraints are applied at these points.

GRID 1000, strictly speaking, is not part of the finite element model. It is used only as a reference point to define Plane 1 for BAR elements 1 through 4. This point is located on the x-z plane since its y-component is 0. Notice also that this point has all its degrees of freedom constrained (see field 8 of the GRID card). This is necessary to prevent the stiffness matrix of the structure from becoming singular.

The square section beam (beam in the x-direction) is modeled with BARs 1, 2, 3, and 4. The orientation of Plane 1 is defined without using the V-vector. Instead, an auxiliary GRID point called G0 is used. The ID of this point (GRID 1000 in this example) goes in field 6 of the CBAR card. With this approach, Plane 1 is defined with two vectors: the vector which goes from one end of the beam to the other, (GA, GB) in this case, and the vector (GA, G0). In this example Plane 1 is the x-z plane. The same would have been achieved by specifying, for example, V = (456.00, 0.0, 99.0).

The two BAR elements used to model the I beam (beam in the z-direction) are defined with the CBAR 5 and CBAR 6 cards. In this case Plane 1 is the x-z plane since the V-vector is (1.0, 0.0, 0.0).

CBAR card number 6 also specifies that there is a hinge at one end of BAR element 6. The entry 456 in the second field of the continuation card states that the connection between this BAR and the rest of the structure transmits neither bending nor torsional moments. In summary, the pin flag specifies the degrees of freedom (expressed in the BAR's coordinate system) that are released at the connection between the BAR end and the grid point. Note that the BAR must have stiffness associated with the degree of freedom released by the pin flag.

Since the two beams have different cross sections, and therefore different properties, two PBAR cards are required, one for each beam. I_1, I_2, and J all need to be specified since the beams are able to deform in any direction (this is actually a three-dimensional problem). If we fail to specify for a certain BAR, the value of J for example, that beam will have no torsional stiffness, which in turn will cause the stiffness matrix to be singular. This, of course, would prevent MSC/NASTRAN from running.

All the loads acting on this structure are prescribed by cards having the same ID (555555). FORCE and MOMENT cards have been explained in previous examples.

The PLOAD1 card used to define the distributed load on BAR 2, demonstrates a different option than that demonstrated in Problem 7: the length (LE) option. The LE that appears in field 5 indicates that X_1 and X_2 are actual distances measured from GA along the beam. We take $X_1 = 0$ and $X_2 = 50$ simply because the length of the BAR is 50 in and the linearly decreasing distributed load is applied over the entire span of the BAR.

9.5 Results

EPSILON appears in Exhibit 9.5 and the displacement vector in Exhibit 9.6. The displacements are reasonable. GRIDs 2 and 4, for instance, present the same displacement in the x-direction and the same rotation about x. This is consistent with the geometry of the model and the loading pattern. GRID 3 exhibits a very small displacement in the y- and z-directions when compared to GRIDs 2 and 4. This is due to the stiffening effect of the I beam.

Exhibit 9.7 shows the reactions at the supports. As expected, there is no torsional moment at GRID 51, because the hinge does not allow torsion to be transmitted to BAR 6.

The output generated by ELFORCES = ALL in the Case Control Deck is depicted in Exhibit 9.8. For each BAR element we obtain at both ends (STATION 0.000 and STATION 1.000) the value of the bending moment, shear, axial force and torque. Figure 9.4 helps to interpret these results. Due to the hinge, BAR element 6 at station 0.000 exhibits zero moments.

The first plot (Figure 9.5) shows the model. The second plot (Figure 9.6) shows the deformed and the undeformed structure.

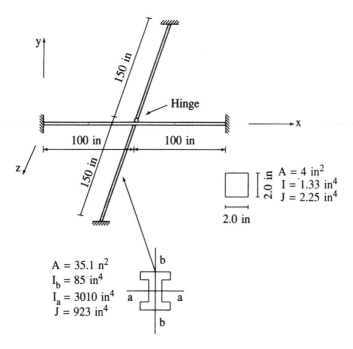

Figure 9.1

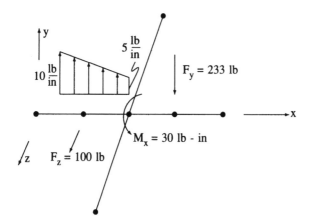

Figure 9.2

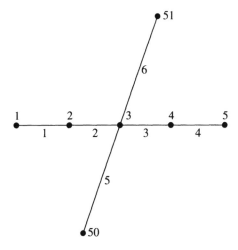

Figure 9.3

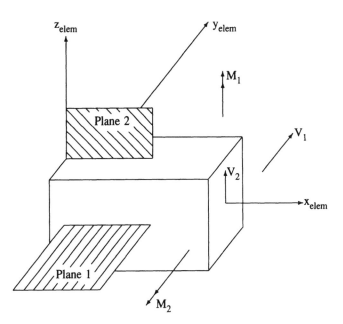

Figure 9.4

```
N A S T R A N   E X E C U T I V E   C O N T R O L   D E C K   E C H O
     ID   ANOTHER,BEAM
     SOL    24
     TIME  4
     CEND
```

Exhibit 9.1   Executive Control Deck


```
            C A S E    C O N T R O L   D E C K   E C H O
     CARD
     COUNT
      1      $
      2      ECHO=BOTH
      3      DISPLACEMENT= ALL
      4      ELFORCES=  ALL
      5      SPCFORCES=   ALL
      6      LOAD  = 555555
      7      $
      8      $$$$  PLOTS  $$$$$
      9      $
     10      OUTPUT(PLOT)
     11      SET 2  INCLUDE   ALL
     12      AXES   Z, X, Y
     13      VIEW   0.0 ,  25., 30.00
     14      FIND
     15      PLOT  LABEL  BOTH
     16      PLOT  STATIC  DEFORMATION  0  SET   2
     17      BEGIN BULK
```

Exhibit 9.2   Case Control Deck

```
.    1  ..    2  ..    3  ..    4  ..    5  ..    6  ..    7  ..    8  ..    9  ..   10  .
         I N P U T   B U L K   D A T A   D E C K   E C H O
$
$$$   GRID POINTS
$
GRID     1                        0.0       .0        .0                 123456
GRID     2                        50.0      .0        .0
GRID     3                        100.0     .0        .0
GRID     4                        150.00    .0        .0
GRID     5                        200.00    .0        .0                 123456
GRID     50                       100.00    .0        150.0              123456
GRID     51                       100.00    .00      -150.0              123456
GRID     1000                     500.      .0        500.               123456
$
$$$$$ BEAMS
$
CBAR     1        100     1        2        1000
CBAR     2        100     2        3        1000
CBAR     3        100     3        4        1000
CBAR     4        100     4        5        1000
CBAR     5        200     3        50       1000                                              +N
+N                                          1.0       .0        .0
CBAR     6        200     3        51       1000                                              +MC
+MC      456                                1.0       .0        .0
```

Exhibit 9.3 Input Bulk Data Deck

```
$$$$
PBAR    100    1    4.      1.33     1.33     2.25
PBAR    200    1    35.1    85.00    3010.00  923.00
MAT1    1    30.0+6    .333
$$$$ LOADS  ( AND MOMENTS ) APPLIED
FORCE   555555    4    233.    0.0    -1.    .0
FORCE   555555    2    100.    -.0     .0   1.0
MOMENT  555555    3    30.0    1.      .0    0.
PLOAD1  555555    2    FY      LE    0.0   10.   50.0   5.
ENDDATA
```

Exhibit 9.3 (continued)   Input Bulk Data Deck

```
                              S O R T E D   B U L K   D A T A   E C H O
CARD    .   1   ..   2   ..   3  ..   4   ..   5   ..   6  ..   7  ..   8   ..   9   ..  10   .

 1-    CBAR     1     100         1        2          1000
 2-    CBAR     2     100         2        3          1000
 3-    CBAR     3     100         3        4          1000
 4-    CBAR     4     100         4        5          1000
 5-    CBAR     5     200         3       50          1.0      .0       .0                        +N
 6-    +N
 7-    CBAR     6     200         3       51          1.0      .0       .0                        +MC
 8-    +MC   456
 9-    FORCE  555555    2               100.         -.0      .0       1.0
10-    FORCE  555555    4               233.          0.0     0.0     -1.       .0
11-    GRID     1               0.0       .0          .0                       123456
12-    GRID     2               50.0      .0          .0
13-    GRID     3               100.0     .0          .0
14-    GRID     4               150.00    .0          .0
15-    GRID     5               200.00    .0          .0                       123456
16-    GRID    50               100.00    .0          150.0                    123456
17-    GRID    51               100.00    .00        -150.0                    123456
18-    GRID   1000              500.      .0          500.                     123456
19-    MAT1     1    30.0+6               .333         1.
20-    MOMENT 555555    3               30.0          .0       .0       0.
21-    PBAR   100       1               4.            1.33     1.33     2.25
22-    PBAR   200       1               35.1          85.00    1.33     3010.00  923.00
23-    PLOAD1 555555    2               FY    LE      0.0      0.0      10.      50.0     5.
       ENDDATA
```

Exhibit 9.4   Bulk Data Deck sorted by alphabetical order

*** USER INFORMATION MESSAGE 5293 FOR DATA BLOCK KLL

LOAD SEQ. NO.	EPSILON	EXTERNAL WORK	EPSILONS LARGER THAN 0.001 ARE FLAGGED WITH ASTERISKS
1	5.1921957E-18	7.8519030E+00	

Exhibit 9.5 Epsilon

			D I S P L A C E M E N T	V E C T O R			
POINT ID.	TYPE	T1	T2	T3	R1	R2	R3
1	G	0.0	0.0	0.0	0.0	0.0	0.0
2	G	-2.482401E-06	2.957273E-02	1.327532E-02	5.500560E-06	-4.454809E-06	3.032472E-04
3	G	-4.964801E-06	1.099098E-03	3.561011E-06	1.100112E-05	1.760557E-05	-1.027655E-04
4	G	-2.482401E-06	-3.114960E-02	-2.182892E-04	5.500560E-06	-4.347979E-06	9.204919E-06
5	G	0.0	0.0	0.0	0.0	0.0	0.0
50	G	0.0	0.0	0.0	0.0	0.0	0.0
51	G	0.0	0.0	0.0	0.0	0.0	0.0
1000	G	0.0	0.0	0.0	0.0	0.0	0.0

Exhibit 9.6 Displacement vector

		F O R C E S	O F	S I N G L E - P O I N T	C O N S T R A I N T		
POINT ID.	TYPE	T1	T2	T3	R1	R2	R3
1	G	5.957761E+00	-8.423646E+01	-5.042318E+01	-2.785355E+00	1.264135E+03	-2.347903E+03
5	G	5.957761E+00	1.184340E-02	4.197724E-01	-2.785355E+00	1.396400E+01	-2.968195E+03
50	G	-1.192678E+01	-8.797661E+01	-2.499829E+01	-1.322092E+04	5.952134E+02	7.115723E+03
51	G	1.125355E-02	-8.822090E+01	-2.499829E+01	1.323313E+04	1.688032E+00	0.0

Exhibit 9.7 Single point constraint forces

F O R C E   D I S T R I B U T I O N   I N   B A R   E L E M E N T S   (C B A R)

ELEMENT ID.	STATION (PCT)	BEND-MOMENT PLANE 1	BEND-MOMENT PLANE 2	SHEAR FORCE PLANE 1	SHEAR FORCE PLANE 2	AXIAL FORCE	TORQUE
1	0.000	1.264135E+03	-2.347903E+03	5.042318E+01	-8.423646E+01	-5.957761E+00	2.785355E+00
1	1.000	-1.257025E+03	1.863920E+03	5.042318E+01	-8.423646E+01	-5.957761E+00	2.785355E+00
2	0.000	-1.257025E+03	1.863920E+03	-4.957682E+01	-8.423645E+01	-5.957761E+00	2.785355E+00
2	1.000	1.221816E+03	-4.340924E+03	-4.957682E+01	2.907635E+02	-5.957761E+00	2.785355E+00
3	0.000	2.801325E+01	2.774799E+03	4.197724E-01	1.145660E+02	5.957761E+00	-2.785355E+00
3	1.000	7.024624E+00	-2.953503E+03	4.197724E-01	1.145660E+02	5.957761E+00	-2.785355E+00
4	0.000	7.024624E+00	-2.953503E+03	4.197724E-01	-1.184340E+02	5.957761E+00	-2.785355E+00
4	1.000	-1.396400E+01	2.968195E+03	4.197724E-01	-1.184340E+02	5.957761E+00	-2.785355E+00
5	0.000	-1.193803E+03	2.442929E+01	-1.192678E+01	-8.797661E+01	-2.499829E+01	7.115722E+03
5	1.000	5.952135E+02	1.322092E+04	-1.192678E+01	-8.797661E+01	-2.499829E+01	7.115722E+03
6	0.000	0.0	0.0	1.125355E-02	8.822090E+01	2.499829E+01	0.0
6	1.000	-1.688032E+00	-1.323313E+04	1.125355E-02	8.822090E+01	2.499829E+01	0.0

Exhibit 9.8   Forces in BAR elements

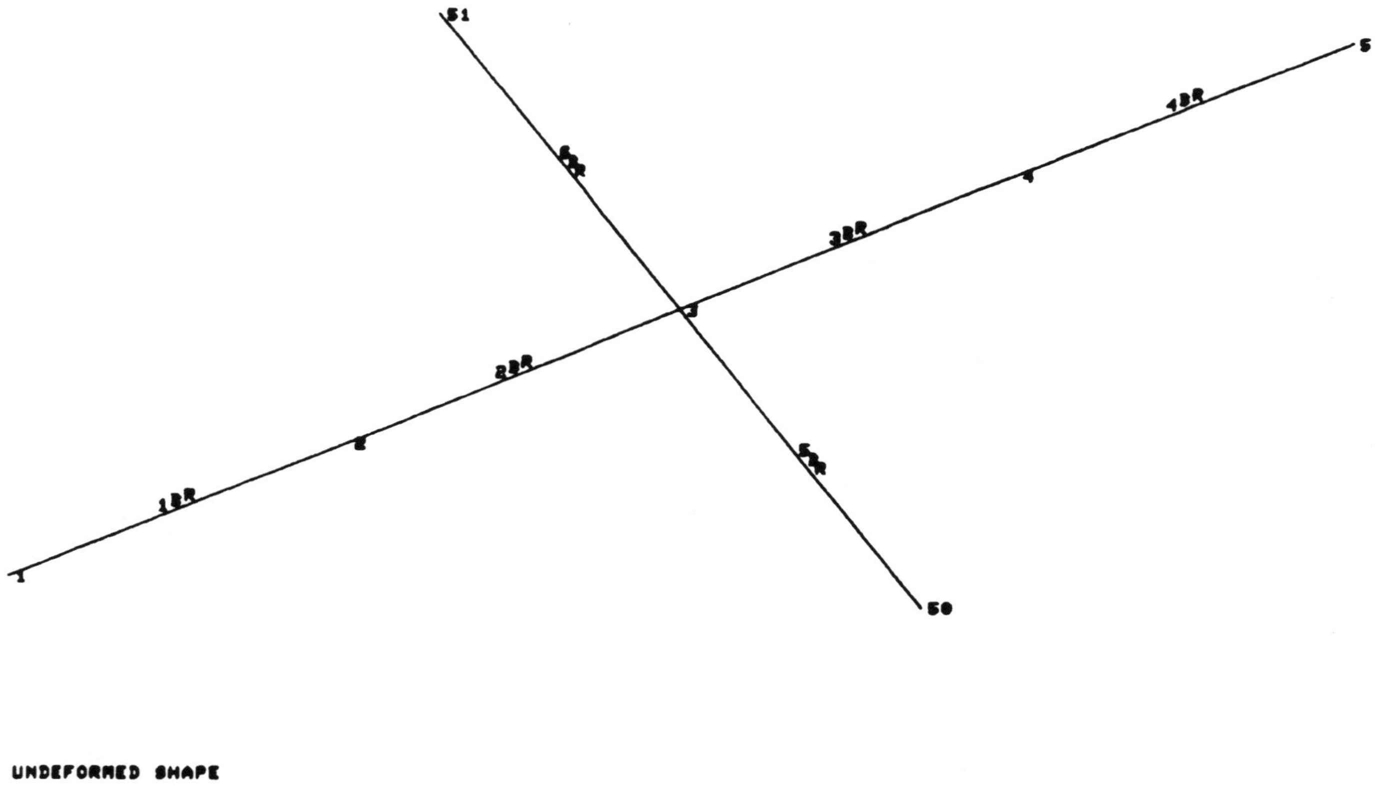
51
5
4BR
4
3BR
3
2BR
2
1BR
1
50
UNDEFORMED SHAPE

Figure 9.5 Finite element model of beam structure using bar elements.

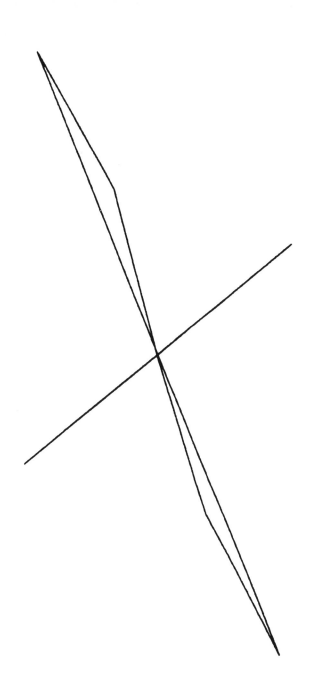

STATIC DEFOR. SUBCASE 1   LOAD 555555

Figure 9.6 Deformed shape of the beam structure.

10

Problem 10

10.1 Statement of the problem

Consider the square plate shown in Figure 10.1. It is supported at the corners on springs that have a stiffness constant equal to 33.0 N/mm and it is uniformly loaded with a distributed load of .01 N/mm^2 in the negative z-direction. Take E = 50,000 N/mm^2 and ν = 0.28. Calculate the displacements at the supports and at the center.

10.2 Cards introduced

Case Control Deck   None

Bulk Data Deck   CELAS2
                                     CQUAD4
                                     PLOAD2
                                     PSHELL

10.3 MSC/NASTRAN formulation

The plate is modeled with four QUAD4 elements. The QUAD4 element is a four-node quadrilateral element that supports bending, membrane, and transverse shear behavior. It can be used to model plates and curved shells.

Plate elements in MSC/NASTRAN do not resist rotations with respect to an axis perpendicular to their surface. Thus, they support only five degrees of freedom per GRID point, three translations and two rotations. The model is outlined in Figure 10.2.

10.4 Input Data Deck

The Executive Control Deck appears in Exhibit 10.1 and the Case Control Deck in Exhibit 10.2.

Examine the Case Control Deck. Notice that DISP = ALL is used instead of DISPLACEMENT = ALL to request that displacements be printed. Both alternatives are equivalent. Many output request cards can be abbreviated for convenience.

The AXES MY, X, Z card (see Figure 1.3) defines R as the negative y-axis. This means that the observer will look at the structure projected onto the x-z plane from the negative y-axis. The choice of VIEW 0.0, 30.0, 25.0 will permit the plate to be seen as a three dimensional structure.

Exhibits 10.3 and 10.4 show the Bulk Data Deck.

The GRID cards show degree of freedom 6 (rotation about z) constrained because the QUAD4 element has no stiffness to prevent rotations in that direction. In addition, the displacements in the x- and y-direction are constrained to prevent the stiffness matrix from becoming singular since the plate is not supported in these directions. To fully describe the response of this plate, only one translational degree of freedom (3), and two rotational degrees of freedom (4 and 5) are necessary.

The connectivity of the plate (QUAD4) elements is determined by CQUAD4 cards. The entry 888 in the third field of the CQUAD4 cards points to a PSHELL card.

The PSHELL card determines the properties of the QUAD4. The third field gives the ID number of the MAT1 card used to supply the properties of the material that provides membrane stiffness. The fifth field is used to give the ID number of the MAT1 card that supplies the properties of the material that resists bending. Since this is a solid homogeneous plate both materials are the same. The effects of the transverse shear flexibility can be included by giving the ID number of the material resisting shear in field 7. This effect, however, is negligible in thin plates and therefore is not included in this example. Field 4 of this card specifies the membrane thickness, 5 mm in this case, which is equal to the plate thickness because this is a solid homogeneous plate.

A bending stiffness parameter goes in field 6. This parameter is 1.0 (default value) for solid homogeneous plates. It defines the ratio between the moment of inertia (per unit of length) of the cross section of the plate which resists bending, and the moment of inertia of the whole cross section.

CELAS2 cards are used to represent the springs connected to the corner GRID points. The ID number of these cards is arbitrary, as long as different springs are defined with cards having different ID numbers. The fourth field contains the ID number of the GRID to which the spring is connected. The fifth field indicates the orientation of the spring, the z-direction (3) in this case. The fact that fields 6 and 7 are blank indicates that the other end the spring is connected to the ground.

The PLOAD2 card (selected by LOAD=1 in the Case Control Deck) indicates that plate (QUAD4) elements 100, 200, 300, and 400 are loaded by a uniform pressure of -.01 N/mm^2 (the positive direction of the pressure is determined according to the right-hand rule using the sequence of grid points G1, G2, G3, and G4 that define the plate element).

10.5 Results

Exhibit 10.5 shows the value of EPSILON and Exhibit 10.6 displays the displacement vector.

The results satisfy the expected symmetry conditions. For example, GRIDs 1, 3, 7, and 9 show the same displacement in the z-direction. The rotations at GRID 5, the center point, should be 0, due to the symmetry. The values obtained (of the order of 10^{-16} and 10^{-17}) can be considered acceptable because they are computational zeros.

Notice also that the deformations in the z-direction are small compared to the thickness of the plate. This is an *essential condition* to ensure that the assumptions of linear thin plate theory are not violated. If deflections of the same order of the thickness of the plate had been obtained the analysis would have not been valid. A new analysis incorporating large deflection geometric nonlinearities would have been required.

The first plot (Figure 10.3) shows the finite element model. The second plot (Figure 10.4), which shows the deformed and the undeformed structure, is consistent with the fact that the plate deflects almost uniformly in the z-direction.

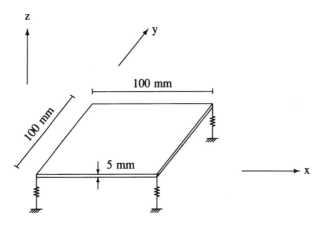

Figure 10.1

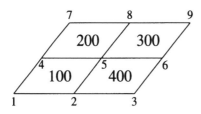

Figure 10.2

```
N A S T R A N   E X E C U T I V E   C O N T R O L   D E C K   E C H O


    ID    PLATE,PROBLEM
    SOL   24
    TIME  3
    CEND
```

Exhibit 10.1   Executive Control Deck

```
        C A S E     C O N T R O L   D E C K   E C H O
 CARD
 COUNT
  1      ECHO=  BOTH
  2      TITLE= PLATE MOUNTED ON SPRINGS
  3      LOAD = 1
  4      DISP=  ALL
  5      $$$
  6      $$$     PLOTS
  7      $$$
  8      OUTPUT(PLOT)
  9      SET  5555   INCLUDE  ALL
 10      AXES  MY, X,  Z
 11      VIEW  0.0,  30.0,  25.0
 12      FIND
 13      PLOT  LABEL  BOTH
 14      PLOT  STATIC DEFORMATION  0   SET  5555
 15      $$$
 16      BEGIN   BULK
```

Exhibit 10.2   Case Control Deck

```
                    I N P U T   B U L K   D A T A   D E C K   E C H O

     .  1  ..  2  ..  3  ..  4  ..  5  ..  6  ..  7  ..  8  ..
     $
     $ GRID POINTS
     $
     $
     GRID    1                 .0      .0      .0              126
     GRID    2               50.0      .0      .0              126
     GRID    3              100.0      .0      .0              126
     GRID    4                 .0    50.0      .0              126
     GRID    5               50.0    50.0      .0              126
     GRID    6              100.00   50.      .0              126
     GRID    7                 .0   100.       .0              126
     GRID    8               50.0   100.       .0              126
     GRID    9              100.0   100.       .0              126
     $
     $    PLATE ELEMENTS
     $
     CQUAD4  100    888      1       2       5       4
     CQUAD4  200    888      4       5       8       7
     CQUAD4  300    888      5       6       9       8
     CQUAD4  400    888      2       3       6       5
     $
     PSHELL  888    4        5.      4       1.0
     MAT1    4      50000.           .28
     $
     $  SPRINGS
     $
     CELAS2  1000   33.      1       3
     CELAS2  1001   33.      3       3
     CELAS2  1002   33.      7       3
     CELAS2  1003   33.      9       3
     $
     $    LOAD (DISTRIBUTED  LOAD)
     $
     PLOAD2  1      -.01     100     200     300     400
     ENDDATA
```

Exhibit 10.3   Input Bulk Data Deck

```
                          S O R T E D   B U L K   D A T A   E C H O
CARD
COUNT     .   1  ..   2  ..   3  ..   4  ..   5  ..   6  ..   7  ..   8
    1-    CELAS2  1000   33.    1       3
    2-    CELAS2  1001   33.    3       3
    3-    CELAS2  1002   33.    7       3
    4-    CELAS2  1003   33.    9       3
    5-    CQUAD4  100    888    1       2       5       4
    6-    CQUAD4  200    888    4       5       8       7
    7-    CQUAD4  300    888    5       6       9       8
    8-    CQUAD4  400    888    2       3       6       5
    9-    GRID    1             .0      .0      .0                    126
   10-    GRID    2             50.0    .0      .0                    126
   11-    GRID    3             100.0   .0      .0                    126
   12-    GRID    4             .0      50.0    .0                    126
   13-    GRID    5             50.0    50.0    .0                    126
   14-    GRID    6             100.00  50.     .0                    126
   15-    GRID    7             .0      100.    .0                    126
   16-    GRID    8             50.0    100.    .0                    126
   17-    GRID    9             100.0   100.    .0                    126
   18-    MAT1    4      50000.          .28
   19-    PLOAD2  1      -.01   100     200     300     400
   20-    PSHELL  888    4      5.      4       1.0
          ENDDATA

          TOTAL COUNT=    21
```

Exhibit 10.4   Bulk Data Deck sorted by alphabetical order

```
*** USER INFORMATION MESSAGE 5293 FOR DATA BLOCK KLL

LOAD SEQ. NO.      EPSILON       EXTERNAL WORK    EPSILONS LARGER THAN 0.001 ARE FLAGGED WITH ASTERISKS
       1        2.0080547E-14    3.8766205E+01
```

Exhibit 10.5 Epsilon

                                    D I S P L A C E M E N T   V E C T O R

POINT ID.	TYPE	T1	T2	T3	R1	R2	R3
1	G	0.0	0.0	-7.575758E-01	-6.123130E-04	6.123130E-04	0.0
2	G	0.0	0.0	-7.776644E-01	-2.516870E-04	2.319515E-17	0.0
3	G	0.0	0.0	-7.575758E-01	-6.123130E-04	-6.123130E-04	0.0
4	G	0.0	0.0	-7.776644E-01	-2.734985E-16	2.516870E-04	0.0
5	G	0.0	0.0	-7.883918E-01	-2.956890E-16	2.935477E-17	0.0
6	G	0.0	0.0	-7.776644E-01	-2.978845E-16	-2.516870E-04	0.0
7	G	0.0	0.0	-7.575758E-01	6.123130E-04	6.123130E-04	0.0
8	G	0.0	0.0	-7.776644E-01	2.516870E-04	4.868068E-17	0.0
9	G	0.0	0.0	-7.575758E-01	6.123130E-04	-6.123130E-04	0.0

Exhibit 10.6 Displacement vector

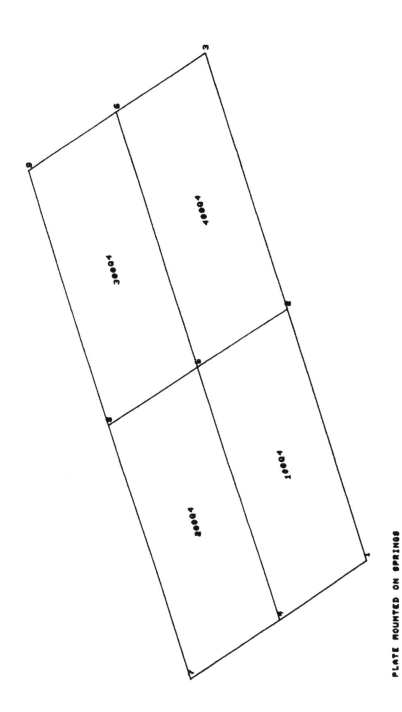

PLATE MOUNTED ON SPRINGS

UNDEFORMED SHAPE

Figure 10.3 Finite element model of the plate using Quad 4 elements.

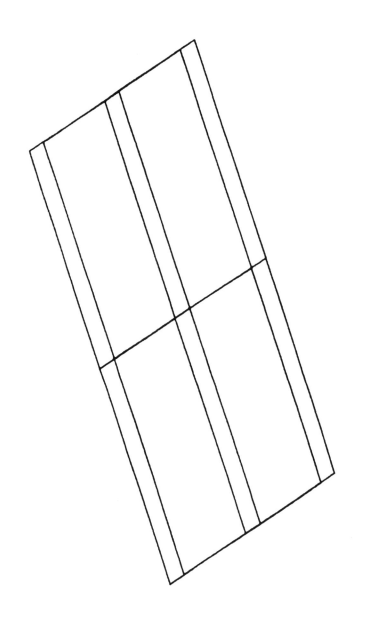

PLATE MOUNTED ON SPRINGS
STATIC DEFOR. SUBCASE 1   LOAD 1

Figure 10.4 Square plate mounted on springs, deformed and undeformed configuration.

11
Problem 11

11.1 Statement of the problem

Consider the plate shown in Figure 11.1. It is clamped on one edge and supported at two points, (x=200 in, y=0 in) and (x=200 in, y=100 in). A load of 100 lb is applied at (x=200 in, y=50 in) in the negative z-direction. Determine the plate forces, stresses, and reactions at the supports. Assume $E = 30 \times 10^6$ psi and $\nu = .33$.

11.2 Cards introduced

<u>Case Control Deck</u>   None

<u>Bulk Data Deck</u>   CTRIA3
Use of replicator cards

11.3 MSC/NASTRAN formulation

The structure is modeled using an array of quadrilateral (QUAD4) and triangular (TRIA3) plate elements as shown in Figure 11.2. Replicator cards are used to take advantage of the regular geometry of certain portions of the structure.

11.4 Input Data Deck

Exhibits 11.1 and 11.2 show the Executive Control Deck and the Case Control Deck, respectively.

Examine the Case Control Deck carefully. The SPC=111 card points to two SPC1 cards in the Bulk Data Deck, which are used to define the boundary conditions at the supports.

The AXES Z, X, Y statement followed by VIEW 0.0, 0.0, 0.0 in the plot subdeck states that the structure is to be projected onto the x-y plane. Recall that PLOT LABEL BOTH creates a plot in which both the GRID points and the finite elements are displayed with their corresponding ID numbers. PLOT SHRINK is a new statement. It is used to ensure that no elements are omitted when creating the model. This will be apparent when examining the plots.

A useful feature of MSC/NASTRAN is that it can create contour plots. These plots show the pattern of variation of a certain recovery variable within the structure. Two contour plots are created in this example. The first shows the maximum shear stress (CONTOUR MAXSHEAR followed by PLOT CONTOUR SET 6666 OUTLINE). The second contour plot displays the displacement in the z-direction (CONTOUR ZDISP followed by PLOT CONTOUR SET 6666 OUTLINE). Other possibilities for contour plots are listed in Appendix IV.

The replicator is a useful feature when a certain pattern repeats itself several times in the model. It can be seen in Figure 11.2 that GRIDs 1, 2, 3, 4, and 5 are equally spaced and located along the same line. Therefore, in order to define these GRIDs, instead of entering one GRID card for each point, the replicator is used to minimize the work.

Consider the GRID cards shown in Exhibit 11.3. The replicator is used as follows. First, we enter the GRID card required to define GRID 1 as usual. The GRID card for GRID 2 is very similar to the GRID card for GRID 1. In fact, it is the same except for the ID number which must be increased by 1 and the x-coordinate which must be increased by 50.0. This is all that needs to be specified on the card immediately following the card for GRID 1. The = symbol on the replicator card means: put in this field the same entry as the corresponding field of the previous card. The *(1) means: put in this field the same entry as the corresponding field of the previous card, plus one. The *(50.) means: put in this field the same entry as the corresponding field of the previous card, plus 50.0. Finally, the == symbol means: fill the remaining fields with whatever the previous card has in these fields. Thus, the card for GRID 2 is generated more efficiently.

The cards for GRIDs 3, 4, and 5 follow the same pattern. Therefore, they can be generated simply by stating that this pattern is to be repeated three more times. This is accomplished by the card that shows a =(3) in the first field. The other GRIDs of the model are generated using the same approach: first, one card is fully described. Then, the second card is defined using the first card as reference. And finally, the number of times this pattern is to be repeated is specified. This approach is not restricted to GRID cards. It can be employed with other cards as well, as is demonstrated with the CQUAD4 cards.

The data deck generated by the replicator is shown in Exhibit 11.4 sorted by alphabetical order.

A few more comments regarding the Bulk Data Deck are pertinent. It is known from the previous problem that plate elements do not support rotations about an axis normal to their plane, the z-axis in this case. For this reason, degree of freedom 6 is constrained on all GRID cards (see field 8).

The boundary (support) conditions are defined using SPC1 cards. These cards are selected by the SPC=111 card in the Case Control Deck[1]. The first SPC1 card states that GRIDs 1, 6, 11, and 16 are clamped, consistent with Figure 11.1. The second SPC1 card constrains the displacement in the z-direction (3) at GRIDs 5 and 15. Both SPC cards and the eighth field of the GRID card are valid alternatives to constrain selected degrees of freedom.

The QUAD4 elements are defined using CQUAD4 cards, which in turn point to a PSHELL card (ID=333) to specify the remaining properties. The only triangular element, TRIA3, is defined using a CTRIA3 card, which also points to the same PSHELL card. This is due to the fact that in this structure all plate elements share the same properties.

The PSHELL card specifies that the thickness of the plate is 1.2 in (fourth field) and that a MAT1 card with ID=1 gives the material properties for both bending and membrane behavior. As in the previous example, a 1.0 is entered in the sixth field. This is the bending parameter for solid homogeneous plates. The continuation card is used to specify the locations at which the plate stresses are to be computed. Since the thickness of the plate is 1.2 in, the choice of 0.6 in and -0.6 in corresponds to the top and bottom fibers.

A FORCE card, selected by LOAD=100 in the Case Control Deck, defines the point load acting on the plate.

11.5 Results

Exhibit 11.5 shows a small EPSILON, a necessary condition to accept the results as valid. Exhibit 11.6 shows the displacements for all GRIDs. The deflections (displacements in the z-direction) are small compared to the thickness of the plate (1.2 in) which means that the assumption of small deflections required by linear thin plate theory holds. There are no displacements in the x- or y-direction. This is to be expected since no forces parallel to the x-y plane act on the plate and no coupling between bending

---

[1] SPC cards in the Case Control Deck can also point to SPC cards in the Bulk Data Deck.

and membrane effects is assumed (MID4 is blank in the PSHELL card.) Exhibit 11.7 shows the reactions at the supports.

Exhibit 11.8 shows the forces in each QUAD4 element. The element internal moments at the centroid of the element (M_x, M_y, M_{xy}) are given in units of moment per unit of length. They are expressed in terms of the element coordinate system of the QUAD4, which is determined by the sequence in which G1, G2, G3, and G4 are specified (see Figure 11.3). The transverse shear forces are also expressed in units of force per unit of length. The output for the TRIA3 element, shown in Exhibit 11.9, can be interpreted in a similar way.

Internal stresses for the QUAD4 elements are presented in Exhibit 11.10, again using the element coordinate system (see Figure 11.3) as a reference. These values correspond to the stresses at the centroid of the element. Note that two sets of values are printed for each element, one at the top surface of the plate (z = thickness/2) and one at the bottom surface of the plate (z = -thickness/2). Stresses for the triangular element are printed in a similar fashion (Exhibit 11.11).

The orientation of the principal axes at the centroid of each element is also given, as well as the major (maximum) and minor (minimum) stresses. The last column corresponds to the so-called Von Mises stress (σ_{VM}) defined as

$$\sigma_{VM} = \{1/2[(\sigma_{major} - \sigma_{minor})^2 + \sigma_{major}^2 + \sigma_{minor}^2]\}^{1/2} \qquad (11.1)$$

Consider as an example the top fiber of element 10. In this case, Eq. 11.1 becomes

$$2.561253 = \{1/2[(2.229997 + 0.5673667)^2 + 2.229997^2 + 0.5673667^2]\}^{1/2} \qquad (11.2)$$

σ_{VM} can also be expressed in terms of σ_x, σ_y, and σ_{xy} as

$$\sigma_{VM} = \{1/2[(\sigma_x - \sigma_y)^2 + \sigma_y^2 + \sigma_x^2] + 3\sigma_{xy}^2\}^{1/2} \qquad (11.3)$$

which, in this case, becomes

$$2.561253 = \{1/2[(2.105814 + 0.4431837)^2 + 0.4431837^2 + $$

$$+ 2.105814^2] + 3 \times 0.5761628^2\}^{1/2} \qquad (11.4)$$

The first plot (Figure 11.4) is generated by PLOT LABEL BOTH. The second plot (PLOT SHRINK) shows the finite element model in a way that allows to check if there are missing elements in the model (Figure 11.5).

The next plot (Figure 11.6) shows the general pattern of behavior for the maximum shear stress on the plate. Ten contour lines are presented. They

correspond to the stress values listed in Exhibit 11.12 (this output appears at
the end of the output file). The contour plot for the deflections is presented
in Figure 11.7. The table with the values represented by the contour lines is
shown in Exhibit 11.13.

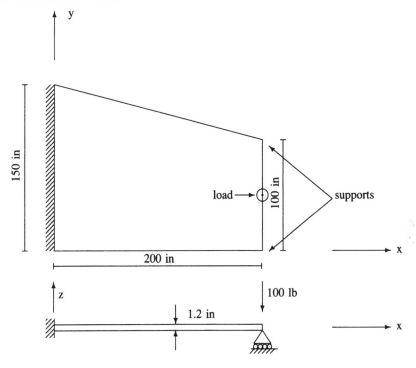

Figure 11.1

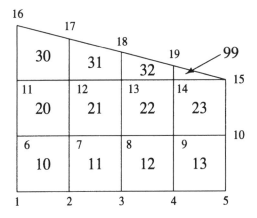

Figure 11.2

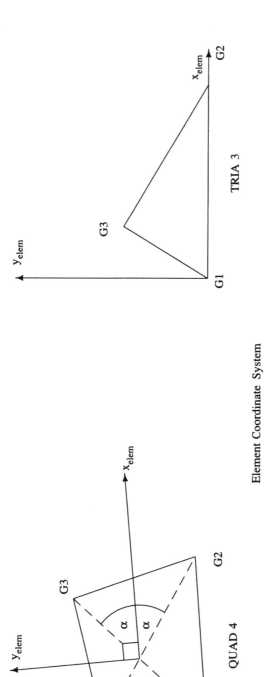

Element Coordinate System

Figure 11.3a

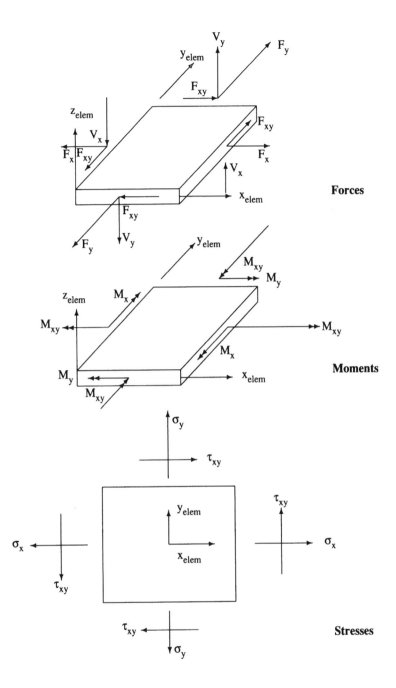

Forces

Moments

Stresses

Figure 11.3b

```
N A S T R A N   E X E C U T I V E   C O N T R O L   D E C K   E C H O

    ID   ANOTHER,EXAMPLE-WITH-A-PLATE
    SOL   24
    TIME  4
    CEND
```

Exhibit 11.1   Executive Control Deck

```
                    C A S E   C O N T R O L   D E C K   E C H O
CARD
COUNT
   1     $
   2     ECHO=BOTH
   3     TITLE=  STATIC ANALYSIS OF PLATE  ---- USING REPLICATOR
   4     LOAD =100
   5     ELFORCES  =  ALL
   6     SPCFORCES =  ALL
   7     STRESS    =  ALL
   8     SPC= 111  $ TO PUT   BOUNDARY CONDITIONS
   9     DISP=  ALL
  10     $$$$$$
  11     $$$$$$   NOW  THE PLOTS
  12     $$$$$$
  13     OUTPUT(PLOT)
  14     SET  6666   INCLUDE   ALL
  15     AXES  Z,  X,  Y
  16     VIEW  0.0,  0.0,  0.0
  17     FIND
  18     PLOT  LABEL  BOTH
  19     PLOT  SHRINK
  20     CONTOUR  MAXSHEAR
  21     PLOT  CONTOUR  SET  6666  OUTLINE
  22     CONTOUR  ZDISP
  23     PLOT  CONTOUR  SET 6666  OUTLINE
  24     $$
  25     BEGIN BULK
```

Exhibit 11.2   Case Control Deck

```
         I N P U T   B U L K   D A T A   D E C K   E C H O

     .  1 ..  2 ..  3 ..  4 ..  5 ..  6 ..  7 ..  8 ..  9 .. 10 .
$$
$$ GRID POINTS USING REPLICATOR
$$
GRID    1          .0       .0       .0                  6
=    *(1)     =  *(50.)      ==
=(3)
$$
GRID    6          .0      50.       .0                  6
=    *(1)     =  *(50.)      ==
=(3)
$$
GRID   11          .0     100.       .0                  6
=    *(1)     =  *(50.)      ==
=(3)
$$
GRID   16          .0     150.       .0                  6
GRID   17        50.0     137.7      .0                  6
GRID   18       100.      125.       .0                  6
GRID   19       150.00    112.5      .0                  6
```

Exhibit 11.3 Input Bulk Data Deck

```
.   1  ..  2  ..  3  ..  4  ..  5  ..  6  ..  7  ..  8  ..  9  ..  10  .
$
$   PLATE ELEMENTS
CQUAD4  10    333    1      2      7      6
=       *(1)  =      *(1)   *(1)   *(1)   *(1)
=(2)
$
CQUAD4  20    333    6      7      12     11
=       *(1)  =      *(1)   *(1)   *(1)   *(1)
=(2)
$
CQUAD4  30    333    11     12     17     16
=       *(1)  =      *(1)   *(1)   *(1)   *(1)
=(1)
CTRIA3  99    333    14     15     19
$$
PSHELL  333   1      1.2    1      1.0                              +H
+H      .6    -.6
MAT1    1     30.+6  .333
$$$
$$$ BOUNDARY CONDITIONS
SPC1    111   123456 1      6      11     16
SPC1    111   3      5      15
$
FORCE   100   10            100.   0.     .0    -1.
ENDDATA
```

Exhibit 11.3 (continued) Input Bulk Data Deck

S O R T E D   B U L K   D A T A   E C H O

CARD COUNT	. 1	.. 2	.. 3	.. 4	.. 5	.. 6	.. 7	.. 8	.. 9	.. 10 .
1-	CQUAD4	10	333	1	2	7	6			
2-	CQUAD4	11	333	2	3	8	7			
3-	CQUAD4	12	333	3	4	9	8			
4-	CQUAD4	13	333	4	5	10	9			
5-	CQUAD4	20	333	6	7	12	11			
6-	CQUAD4	21	333	7	8	13	12			
7-	CQUAD4	22	333	8	9	14	13			
8-	CQUAD4	23	333	9	10	15	14			
9-	CQUAD4	30	333	11	12	17	16			
10-	CQUAD4	31	333	12	13	18	17			
11-	CQUAD4	32	333	13	14	19	18			
12-	CTRIA3	99	333	14	15	19				
13-	FORCE	100	10	100.	0.	.0	-1.			
14-	GRID	1	.0	.0	0.		6			
15-	GRID	2	50.	.0	.0		6			
16-	GRID	3	100.	.0	.0		6			
17-	GRID	4	150.	.0	.0		6			
18-	GRID	5	200.	.0	.0		6			
19-	GRID	6	.0	50.	.0		6			
20-	GRID	7	50.	50.	.0		6			
21-	GRID	8	100.	50.	.0		6			
22-	GRID	9	150.	50.	.0		6			
23-	GRID	10	200.	50.	.0		6			

Exhibit 11.4 Bulk Data Deck sorted by alphabetical order

24-	GRID	11	.0	100.	.0	6
25-	GRID	12	50.	100.	.0	6
26-	GRID	13	100.	100.	.0	6
27-	GRID	14	150.	100.	.0	6
28-	GRID	15	200.	100.	.0	6
29-	GRID	16	.0	150.	.0	6
30-	GRID	17	50.0	137.7	.0	6
31-	GRID	18	100.	125.	.0	6
32-	GRID	19	150.00	112.5	.0	6
33-	MAT1	1	30.+6	.333		
34-	PSHELL	333	1	1	1.0	+H
35-	+H	.6	-.6			
36-	SPC1	111	3	5	15	
37-	SPC1	111	123456	1	6	11	16
	ENDDATA					

Exhibit 11.4 (continued) Bulk Data Deck sorted by alphabetical order

*** USER INFORMATION MESSAGE 5293 FOR DATA BLOCK KLL

LOAD SEQ. NO.	EPSILON	EXTERNAL WORK	EPSILONS LARGER THAN 0.001 ARE FLAGGED WITH ASTERISKS
1	-5.4842683E-16	4.7516772E-01	

Exhibit 11.5 Epsilon

D I S P L A C E M E N T   V E C T O R

POINT ID.	TYPE	T1	T2	T3	R1	R2	R3
1	G	0.0	0.0	0.0	0.0	0.0	0.0
2	G	0.0	0.0	-7.282419E-05	-3.818601E-06	2.632280E-06	0.0
3	G	0.0	0.0	-3.054516E-04	-7.026904E-06	6.518650E-06	0.0
4	G	0.0	0.0	-7.230399E-05	-9.235115E-05	-1.635120E-05	0.0
5	G	0.0	0.0	0.0	-3.464308E-04	1.625370E-05	0.0
6	G	0.0	0.0	0.0	0.0	0.0	0.0
7	G	0.0	0.0	-2.536946E-04	2.539285E-06	9.886574E-06	0.0
8	G	0.0	0.0	-8.430673E-04	-9.952518E-07	1.321411E-05	0.0
9	G	0.0	0.0	-2.955314E-03	2.995935E-06	7.097499E-05	0.0
10	G	0.0	0.0	-9.503354E-03	6.518882E-06	1.878357E-04	0.0
11	G	0.0	0.0	0.0	0.0	0.0	0.0
12	G	0.0	0.0	1.125361E-04	6.079717E-06	-4.456903E-06	0.0
13	G	0.0	0.0	7.549448E-05	2.422717E-05	6.227180E-06	0.0
14	G	0.0	0.0	1.382477E-05	8.856303E-05	-5.133192E-06	0.0
15	G	0.0	0.0	0.0	3.313173E-04	8.555598E-06	0.0
16	G	0.0	0.0	0.0	0.0	0.0	0.0
17	G	0.0	0.0	2.403453E-04	2.528482E-07	-9.131718E-06	0.0
18	G	0.0	0.0	7.884213E-04	3.023655E-05	-2.065256E-05	0.0
19	G	0.0	0.0	1.118211E-03	7.930465E-05	-2.146059E-05	0.0

Exhibit 11.6   Displacement vector

FORCES OF SINGLE-POINT CONSTRAINT

POINT ID.	TYPE	T1	T2	T3	R1	R2	R3
1	G	0.0	0.0	-1.058966E+01	-2.608564E+02	-1.215739E+01	0.0
5	G	0.0	0.0	4.878808E+01	0.0	0.0	0.0
6	G	0.0	0.0	2.221430E+01	-4.594546E+00	-4.212338E+01	0.0
11	G	0.0	0.0	-9.745239E+00	2.666022E+02	-2.142261E-01	0.0
15	G	0.0	0.0	5.115499E+01	0.0	0.0	0.0
16	G	0.0	0.0	-1.822469E+00	2.052874E+01	4.311014E+01	0.0

Exhibit 11.7 Single point constraint forces

FORCES IN QUADRILATERAL ELEMENTS (QUAD4)

ELEMENT ID	- MEMBRANE FORCES -			- BENDING MOMENTS -			- TRANSVERSE SHEAR FORCES -	
	FX	FY	FXY	MX	MY	MXY	QX	QY
10	0.0	0.0	0.0	-5.053953E-01	1.063641E-01	-1.382791E-01	-3.159562E-03	-3.472381E-02
11	0.0	0.0	0.0	-1.500487E-01	4.852619E-01	-3.353040E-01	-3.663830E-03	-1.133340E-01
12	0.0	0.0	0.0	-5.499655E-02	4.361248E+00	-2.841457E+00	-4.707802E-03	-2.300014E-01
13	0.0	0.0	0.0	-8.880910E-03	1.936346E+01	-8.255395E+00	-1.844127E-03	-3.859091E-01
20	0.0	0.0	0.0	-2.065328E-01	8.417124E-02	3.720852E-01	-1.263312E-03	3.515949E-02
21	0.0	0.0	0.0	-2.154188E-01	1.170821E+00	5.824276E-01	-1.082173E-03	1.139333E-01
22	0.0	0.0	0.0	-4.619535E-01	4.632277E+00	2.453653E+00	-9.757927E-03	2.374032E-01
23	0.0	0.0	0.0	2.964943E-01	1.782652E+01	8.129013E+00	-1.408207E-03	4.052497E-01
30	0.0	0.0	0.0	5.569085E-01	-6.466316E-02	2.210540E-01	3.622579E-03	6.700056E-03
31	0.0	0.0	0.0	6.453329E-02	4.224720E-01	1.589921E+00	9.943587E-04	1.735057E-02
32	0.0	0.0	0.0	3.374025E-01	8.223773E-01	3.723126E-01	1.806854E-02	3.483854E-01

Exhibit 11.8 Forces in quadrilateral elements (QUAD4)

```
                F O R C E S   I N   T R I A N G U L A R   E L E M E N T S   ( T R I A 3 )

ELEMENT      - MEMBRANE FORCES -              - BENDING MOMENTS -              - TRANSVERSE SHEAR FORCES -
  ID      FX        FY        FXY        MX          MY          MXY             QX            QY
  99     0.0       0.0        0.0    -2.528605E+00 -4.041720E+00  9.983766E+00  1.691078E-02 -1.946866E+00
```

Exhibit 11.9   Forces in triangular elements (TRIA3)

```
                  S T R E S S E S   I N   Q U A D R I L A T E R A L   E L E M E N T S   ( Q U A D 4 )
ELEMENT   FIBRE         STRESSES IN ELEMENT COORD SYSTEM          PRINCIPAL STRESSES (ZERO SHEAR)
  ID.    DISTANCE     NORMAL-X      NORMAL-Y     SHEAR-XY      ANGLE      MAJOR         MINOR        VON MISES
  10   6.000000E-01  2.105814E+00 -4.431837E-01  5.761628E-01  12.1632  2.229997E+00 -5.673667E-01  2.561253E+00
      -6.000000E-01 -2.105814E+00  4.431837E-01 -5.761628E-01 -77.8368  5.673667E-01 -2.229997E+00  2.561253E+00
  11   6.000000E-01  6.252031E-01 -2.021924E+00  1.397100E+00  23.2741  1.226142E+00 -2.622863E+00  3.405706E+00
      -6.000000E-01 -6.252031E-01  2.021924E+00 -1.397100E+00 -66.7259  2.622863E+00 -1.226142E+00  3.405706E+00
  12   6.000000E-01  2.291522E-01 -1.817186E+01  1.183940E+01  26.0744  6.022670E+00 -2.396538E+01  2.747631E+01
      -6.000000E-01 -2.291522E-01  1.817186E+01 -1.183940E+01 -63.9256  2.396538E+01 -6.022670E+00  2.747631E+01
  13   6.000000E-01  3.700379E-02 -8.068109E+01  3.439748E+01  20.2203  1.270662E+01 -9.335070E+01  1.003094E+02
      -6.000000E-01 -3.700379E-02  8.068109E+01 -3.439748E+01 -69.7797  9.335070E+01 -1.270662E+01  1.003094E+02
  20   6.000000E-01  8.605533E-01 -3.507135E-01 -1.550355E+00 -34.3312  1.919369E+00 -1.409530E+00  2.894160E+00
      -6.000000E-01 -8.605533E-01  3.507135E-01  1.550355E+00  55.6688  1.409530E+00 -1.919369E+00  2.894160E+00
```

Exhibit 11.10   Stresses in quadrilateral elements (QUAD4)

ELEMENT ID.	FIBRE DISTANCE	NORMAL-X	NORMAL-Y	SHEAR-XY	ANGLE	MAJOR	MINOR	VON MISES
21	6.000000E-01	8.975782E-01	-4.878421E+00	-2.426782E+00	-20.0201	1.781821E+00	-5.762663E+00	6.830169E+00
	-6.000000E-01	-8.975782E-01	4.878421E+00	2.426782E+00	69.9799	5.762663E+00	-1.781821E+00	6.830169E+00
22	6.000000E-01	1.924806E+00	-1.930115E+01	-1.023355E+01	-21.9646	6.048052E+00	-2.342440E+01	2.696207E+01
	-6.000000E-01	-1.924806E+00	1.930115E+01	1.023355E+01	68.0354	2.342440E+01	-6.048052E+00	2.696207E+01
23	6.000000E-01	-1.235393E+00	-7.427718E+01	-3.387089E+01	-21.4220	1.205348E+01	-8.756605E+01	9.417311E+01
	-6.000000E-01	1.235393E+00	7.427718E+01	3.387089E+01	68.5780	8.756605E+01	-1.205348E+01	9.417311E+01
30	6.000000E-01	-2.320452E+00	2.694298E-01	-9.210584E-01	-72.2884	5.635831E-01	-2.614605E+00	2.937232E+00
	-6.000000E-01	2.320452E+00	-2.694298E-01	9.210584E-01	17.7116	2.614605E+00	-5.635831E-01	2.937232E+00
31	6.000000E-01	-2.688887E-01	-1.760300E+00	-6.624668E+00	-41.7888	5.651912E+00	-7.681100E+00	1.159122E+01
	-6.000000E-01	2.688887E-01	1.760300E+00	6.624668E+00	48.2112	7.681100E+00	-5.651912E+00	1.159122E+01
32	6.000000E-01	-1.405844E+00	-3.426572E+00	-1.551302E+01	-43.1368	1.312968E+01	-1.796210E+01	2.703446E+01
	-6.000000E-01	1.405844E+00	3.426572E+00	1.551302E+01	46.8632	1.796210E+01	-1.312968E+01	2.703446E+01

Exhibit 11.10 (continued)   Stresses in quadrilateral elements (QUAD4)

STRESSES IN TRIANGULAR ELEMENTS (TRIA 3)

ELEMENT ID.	FIBRE DISTANCE	STRESSES IN ELEMENT COORD SYSTEM				PRINCIPAL STRESSES (ZERO SHEAR)		VON MISES
		NORMAL-X	NORMAL-Y	SHEAR-XY	ANGLE	MAJOR	MINOR	
99	6.000000E-01	1.053585E+01	1.684050E+01	-4.159902E+01	-47.1668	5.540647E+01	-2.803011E+01	7.354327E+01
	-6.000000E-01	-1.053585E+01	-1.684050E+01	4.159902E+01	42.8332	2.803011E+01	-5.540647E+01	7.354327E+01

Exhibit 11.11   Stresses in triangular elements (TRIA3)

MESSAGES FROM THE PLOT MODULE

C O N T O U R   P L O T T I N G   D A T A

ABOVE PLOT IS A CONTOUR PLOT OF SHEAR - MAXIMUM   MIN = 2.561253E+00   MAX = 1.003094E+02

THE CONTOUR VALUES ARE CALCULATED AT FIBER DISTANCE Z1

TABLE OF PLOTTING SYMBOLS

SYMBOL	VALUE	SYMBOL	VALUE	SYMBOL	VALUE	SYMBOL	VALUE
1	2.561253E+00						
2	1.342216E+01						
3	2.428307E+01						
4	3.514398E+01						
5	4.600489E+01						
6	5.686580E+01						
7	6.772671E+01						
8	7.858762E+01						
9	8.944852E+01						
10	1.003094E+02						

Exhibit 11.12   Table of plotting symbols

```
C O N T O U R   P L O T T I N G   D A T A            MESSAGES FROM THE PLOT MODULE

   ABOVE PLOT IS A CONTOUR PLOT OF DEFORMATION Z      MIN = -9.503354E-03      MAX = 1.118211E-03
   IN A COMMON SYSTEM.

                                            TABLE OF   PLOTTING   SYMBOLS
                   SYMBOL  VALUE              SYMBOL  VALUE         SYMBOL  VALUE         SYMBOL  VALUE

SYMBOL  VALUE
   1  -9.503354E-03
   2  -8.323180E-03
   3  -7.143007E-03
   4  -5.962833E-03
   5  -4.782659E-03
   6  -3.602485E-03
   7  -2.422311E-03
   8  -1.242137E-03
   9  -6.196345E-05
  10   1.118211E-03
```

Exhibit 11.13   Table of plotting symbols

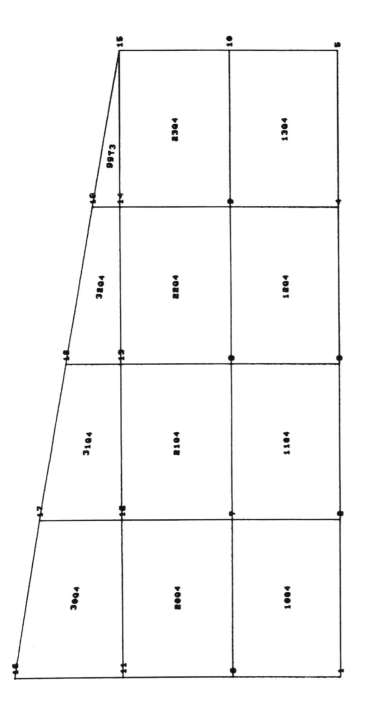

STATIC ANALYSIS OF PLATE ---- USING REPLICATOR

UNDEFORMED SHAPE

Figure 11.4 Finite element model of the plate.

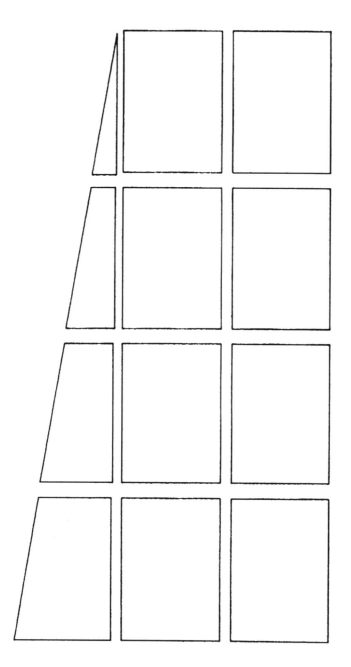

STATIC ANALYSIS OF PLATE ---- USING REPLICATOR

UNDEFORMED SHAPE

Figure 11.5 Finite element model of the plate generated with the "plot shrink" option.

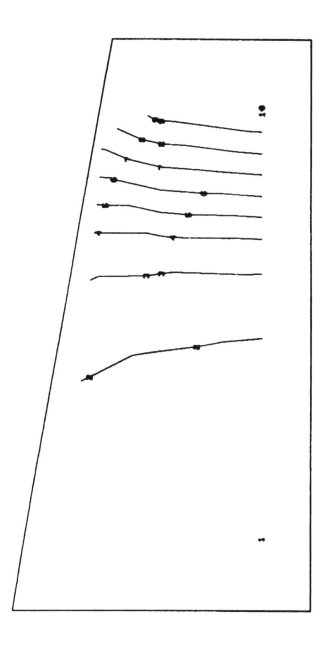

STATIC ANALYSIS OF PLATE ---- USING REPLICATOR

STATIC STRESS SUBCASE 1   LOAD 100

Figure 11.6 Contour plot for maximum shear in the plate.

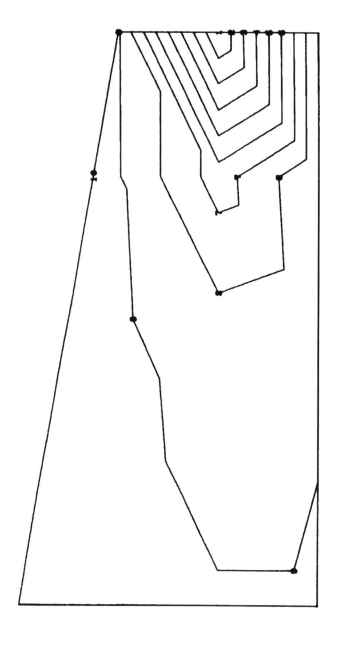

STATIC ANALYSIS OF PLATE ---- USING REPLICATOR

STATIC DEFOR. SUBCASE 1   LOAD 100

Figure 11.7 Contour plot for plate deflection.

12
Problem 12

12.1 Statement of the problem

Consider a thin steel sheet as shown in Figure 12.1. Determine the stresses on this sheet if the right edge is pulled with a force of 240,000 N distributed uniformly along the edge. Use a value of E equal to 200,000 N/mm^2 and a Poisson's ratio of 0.333.

12.2 Cards introduced

<u>Case Control Deck</u>   None

<u>Bulk Data Deck</u>   CQUAD8

12.3 MSC/NASTRAN formulation

This constitutes a typical plane stress problem, which means that the state of stress can be fully described in terms of σ_x, σ_y, and σ_{xy}. This is because the forces act parallel to the plane of the plate (x-y plane) and are assumed to be uniformly distributed through its thickness.

The structure can be modeled using any of the plate elements provided by MSC/NASTRAN since they all support membrane behavior. Figure 12.2 shows the model, which consists of four QUAD8 elements.

The QUAD8 is similar to the QUAD4 except that it features mid-side nodes. In addition, stresses are recovered not only at the centroid of the element but also at the corner grids.

12.4 Input Data Deck

Exhibit 12.1 shows the Executive Control Deck. Interpretation of the Case Control Deck (Exhibit 12.2) should be straightforward at this stage. The Bulk Data Deck is only printed sorted by alphabetical order (Exhibit 12.3), since the ECHO = BOTH card was omitted in the Case Control Deck.

The QUAD8 elements are defined with CQUAD8 cards. The CQUAD8 card is very similar to the CQUAD4 card except that specifies eight GRIDs (G1, G2,..., G8) instead of four. This card points to a PSHELL card, just like the QUAD4 or TRIA3 cards. The PSHELL card defines the membrane thickness to be 1.0 mm; it also gives the ID number of the MAT1 card that supplies the properties of the material resisting membrane effects. In this problem no bending effects are included on the MAT1 card.

The boundary conditions are specified with SPC1 cards. All six degrees of freedom are suppressed at the supports. The remaining GRID points have only two degrees of freedom, translation in the x- and y-directions, consistent with the plane stress assumption made initially.

The force of 240,000 N must be transferred to the GRIDs located at y = 200 mm. Consistent with the assumption that this force is applied uniformly along that edge, the corner GRIDs take only a half of the force taken by the interior GRIDs. This results in a point load of 30,000 N at GRIDs 13 and 5, and point loads of 60,000 N at GRIDs 8, 18, and 19. FORCE cards are employed to define these loads.

12.5 Results

Exhibit 12.4 shows the value of EPSILON. Exhibit 12.5 shows the membrane forces (force per unit of length) using the element coordinate system as reference. The element coordinate system of the QUAD8 element is analogous to that of the QUAD4 element described in Figure 11.3. The membrane forces are given at the centroid of the element and at the corners. As expected, F_x is much larger than F_y.

Exhibit 12.6 shows the stresses in each element at the centroid and at the corners using the element coordinate system as reference. Interpretation of these values is analogous to the case of a QUAD4 element (Figure 11.3). Notice, however, that in this case the stresses at the top surface (z = thickness/2) and at the bottom surface (z = -thickness/2) are the same. This is characteristic of a plane stress problem in which the forces are uniformly distributed over the membrane cross section. Remember that in a pure bending problem, as in the previous example, the stresses at the top and at the bottom have the same absolute value, but opposite sign.

The plots are shown in Figures 12.3 and 12.4. The deformed shape is in agreement with intuition.

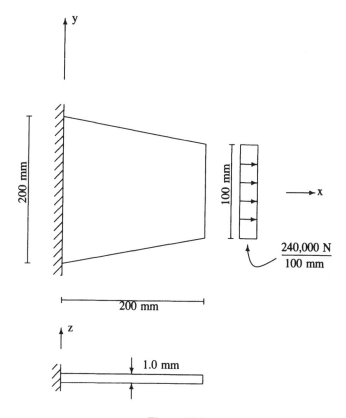

Figure 12.1

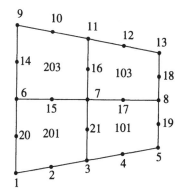

Figure 12.2

```
N A S T R A N   E X E C U T I V E   C O N T R O L   D E C K   E C H O

    ID    PLANE,STRESS
    TIME  2
    SOL 24
    CEND
```

Exhibit 12.1   Executive Control Deck

```
                    C A S E    C O N T R O L   D E C K   E C H O
CARD
COUNT
  1     TITLE=  PLANE STRESS WITH QUAD8'S
  2     LOAD =11
  3     ELFORCES=ALL
  4     SPC=444
  5     STRESS= ALL
  6     $$
  7     $$  PLOTS
  8     $$
  9     OUTPUT(PLOT)
 10     SET   1111   INCLUDE   ALL
 11     AXES  Z,X,Y
 12     VIEW   0.0,  0.0,  0.0
 13     FIND
 14     PLOT   LABEL  BOTH
 15     PLOT   STATIC  DEFORMATION  0  SET  1111
 16     $$
 17     BEGIN BULK

          INPUT BULK DATA CARD COUNT =      48
```

Exhibit 12.2   Case Control Deck

S O R T E D   B U L K   D A T A   E C H O

CARD COUNT	.	1	2	3	4	5	6	7	8	9	10 .
1-	CQUAD8	101	22	3	5	8	7	4	19		+MMM
2-	+MMM	17	21								
3-	CQUAD8	103	22	7	8	13	11	17	18		+BB
4-	+BB	12	16								
5-	CQUAD8	201	22	1	3	7	6	2	21		+VVVV
6-	+VVVV	15	20								
7-	CQUAD8	203	22	6	7	11	9	15	16		+CCDD
8-	+CCDD	10	14								
9-	FORCE	11	5		30000.	1.	.0	.0			
10-	FORCE	11	8		60000.	1.	.0	.0			
11-	FORCE	11	13		30000.	1.	.0	.0			
12-	FORCE	11	18		60000.	1.	.0	.0			
13-	FORCE	11	19		60000.	1.	.0	.0			
14-	GRID	1		0.	-100.	.0					
15-	GRID	2		50.	-87.5	.0					
16-	GRID	3		100.	-75.	.0					
17-	GRID	4		150.	-62.5	.0					
18-	GRID	5		200.	-50.	.0					
19-	GRID	6		.0	.0	.0					
20-	GRID	7		100.	.0	.0					
21-	GRID	8		200.	.0	.0					
22-	GRID	9		.0	100.	.0					

Exhibit 12.3   Bulk Data Deck sorted by alphabetical order

```
23-    GRID      10         50.       87.5      .0
24-    GRID      11         100.      75.       .0
25-    GRID      12         150.      62.5      .0
26-    GRID      13         200.      50.       .0
27-    GRID      14         0.0       50.       .0
28-    GRID      15         50.00     .0        .0
29-    GRID      16         100.      37.5      .0
30-    GRID      17         150.0     .0        .0
31-    GRID      18         200.      25.       .0
32-    GRID      19         200.      -25.0     .0
33-    GRID      20         .0        -50.      .0
34-    GRID      21         100.      -37.5     .0
35-    MAT1      1          200000.   .333
36-    PSHELL    22         1         1.
37-    SPC1      444        3456      2         15        10
38-    SPC1      444        3456      3         21        7     16   11
39-    SPC1      444        3456      4         17        12
40-    SPC1      444        3456      5         19        8     18   13
41-    SPC1      444        123456    1         20        6     14   9
       ENDDATA
```

Exhibit 12.3 (continued) Bulk Data Deck sorted by alphabetical order

*** USER INFORMATION MESSAGE 5293 FOR DATA BLOCK KLL
    EPSILONS LARGER THAN 0.001 ARE FLAGGED WITH ASTERISKS

LOAD SEQ. NO.	EPSILON	EXTERNAL WORK
1	-1.9415656E-17	2.0326434E+05

Exhibit 12.4 Epsilon

ELEMENT		FORCES IN QUADRILATERAL ELEMENTS (QUAD 8)							
		- MEMBRANE FORCES -			- BENDING MOMENTS -			- TRANSVERSE SHEAR FORCES -	
ID	GRID-ID	FX	FY	FXY	MX	MY	MXY	QX	QY
101	CEN/8	1.975018E+03	1.106882E+02	1.043024E+02	0.0	0.0	0.0	0.0	0.0
	3	1.430567E+03	3.630710E+01	3.862249E+02	0.0	0.0	0.0	0.0	0.0
	5	2.229912E+03	5.361137E+01	-2.975563E+02	0.0	0.0	0.0	0.0	0.0
	8	2.477241E+03	5.216447E+02	4.817066E+02	0.0	0.0	0.0	0.0	0.0
	7	1.762351E+03	-1.688104E+02	-1.531657E+02	0.0	0.0	0.0	0.0	0.0
103	CEN/8	1.975018E+03	1.106882E+02	-1.043024E+02	0.0	0.0	0.0	0.0	0.0
	7	1.762351E+03	-1.688104E+02	1.531657E+02	0.0	0.0	0.0	0.0	0.0
	8	2.477241E+03	5.216447E+02	-4.817066E+02	0.0	0.0	0.0	0.0	0.0
	13	2.229912E+03	5.361137E+01	2.975563E+02	0.0	0.0	0.0	0.0	0.0
	11	1.430567E+03	3.630710E+01	-3.862249E+02	0.0	0.0	0.0	0.0	0.0
201	CEN/8	1.399866E+03	8.974188E+01	8.790172E+01	0.0	0.0	0.0	0.0	0.0
	1	1.071818E+03	1.543916E+02	3.017778E+02	0.0	0.0	0.0	0.0	0.0
	3	1.365577E+03	7.245544E-01	1.364605E+02	0.0	0.0	0.0	0.0	0.0
	7	1.812767E+03	-1.769386E+02	-9.809570E+01	0.0	0.0	0.0	0.0	0.0
	6	1.349300E+03	3.807900E+02	1.146431E+01	0.0	0.0	0.0	0.0	0.0
203	CEN/8	1.399866E+03	8.974188E+01	-8.790172E+01	0.0	0.0	0.0	0.0	0.0
	6	1.349300E+03	3.807900E+02	-1.146431E+01	0.0	0.0	0.0	0.0	0.0
	7	1.812767E+03	-1.769386E+02	9.809570E+01	0.0	0.0	0.0	0.0	0.0
	11	1.365577E+03	7.245544E-01	-1.364605E+02	0.0	0.0	0.0	0.0	0.0
	9	1.071818E+03	1.543916E+02	-3.017778E+02	0.0	0.0	0.0	0.0	0.0

Exhibit 12.5 Forces in quadrilateral elements (QUAD8)

S T R E S S E S   I N   Q U A D R I L A T E R A L   E L E M E N T S   (Q U A D   8)

ELEMENT ID	GRID-ID	FIBRE DISTANCE	STRESSES IN ELEMENT COORD SYSTEM			PRINCIPAL STRESSES (ZERO SHEAR)			VON MISES
			NORMAL-X	NORMAL-Y	SHEAR-XY	ANGLE	MAJOR	MINOR	
101	CEN/8	-5.000000E-01	1.975018E+03	1.106882E+02	1.043024E+02	3.1922	1.980835E+03	1.048710E+02	1.930537E+03
		5.000000E-01	1.975018E+03	1.106882E+02	1.043024E+02	3.1922	1.980835E+03	1.048710E+02	1.930537E+03
	3	-5.000000E-01	1.430567E+03	3.630710E+01	3.862249E+02	14.4937	1.530406E+03	-6.353212E+01	1.563141E+03
		5.000000E-01	1.430567E+03	3.630710E+01	3.862249E+02	14.4937	1.530406E+03	-6.353212E+01	1.563141E+03
	5	-5.000000E-01	2.229912E+03	5.361137E+01	-2.975563E+02	-7.6469	2.269863E+03	1.366115E+01	2.263063E+03
		5.000000E-01	2.229912E+03	5.361137E+01	-2.975563E+02	-7.6469	2.269863E+03	1.366115E+01	2.263063E+03
	8	-5.000000E-01	2.477241E+03	5.216447E+02	4.817066E+02	13.1134	2.589457E+03	4.094289E+02	2.410958E+03
		5.000000E-01	2.477241E+03	5.216447E+02	4.817066E+02	13.1134	2.589457E+03	4.094289E+02	2.410958E+03
	7	-5.000000E-01	1.762351E+03	-1.688104E+02	-1.531657E+02	-4.5067	1.774423E+03	-1.808829E+02	1.871432E+03
		5.000000E-01	1.762351E+03	-1.688104E+02	-1.531657E+02	-4.5067	1.774423E+03	-1.808829E+02	1.871432E+03
103	CEN/8	-5.000000E-01	1.975018E+03	1.106882E+02	-1.043024E+02	-3.1922	1.980835E+03	1.048710E+02	1.930537E+03
		5.000000E-01	1.975018E+03	1.106882E+02	-1.043024E+02	-3.1922	1.980835E+03	1.048710E+02	1.930537E+03
	7	-5.000000E-01	1.762351E+03	-1.688104E+02	1.531657E+02	4.5067	1.774423E+03	-1.808829E+02	1.871432E+03
		5.000000E-01	1.762351E+03	-1.688104E+02	1.531657E+02	4.5067	1.774423E+03	-1.808829E+02	1.871432E+03
	8	-5.000000E-01	2.477241E+03	5.216447E+02	-4.817066E+02	-13.1134	2.589457E+03	4.094289E+02	2.410958E+03
		5.000000E-01	2.477241E+03	5.216447E+02	-4.817066E+02	-13.1134	2.589457E+03	4.094289E+02	2.410958E+03
	13	-5.000000E-01	2.229912E+03	5.361137E+01	2.975563E+02	7.6469	2.269863E+03	1.366115E+01	2.263063E+03
		5.000000E-01	2.229912E+03	5.361137E+01	2.975563E+02	7.6469	2.269863E+03	1.366115E+01	2.263063E+03
	11	-5.000000E-01	1.430567E+03	3.630710E+01	-3.862249E+02	-14.4937	1.530406E+03	-6.353212E+01	1.563141E+03
		5.000000E-01	1.430567E+03	3.630710E+01	-3.862249E+02	-14.4937	1.530406E+03	-6.353212E+01	1.563141E+03

Exhibit 12.6   Stresses in quadrilateral elements (QUAD8

201 CEN/8	-5.000000E-01	1.399866E+03	8.974188E+01	8.790172E+01	3.8214	1.405737E+03	8.387050E+01	1.365735E+03
	5.000000E-01	1.399866E+03	8.974188E+01	8.790172E+01	3.8214	1.405737E+03	8.387050E+01	1.365735E+03
1	-5.000000E-01	1.071818E+03	1.543916E+02	3.017778E+02	16.6700	1.162184E+03	6.402594E+01	1.131531E+03
	5.000000E-01	1.071818E+03	1.543916E+02	3.017778E+02	16.6700	1.162184E+03	6.402594E+01	1.131531E+03
3	-5.000000E-01	1.365577E+03	7.245544E-01	1.364605E+02	5.6540	1.379087E+03	-1.278529E+01	1.385524E+03
	5.000000E-01	1.365577E+03	7.245544E-01	1.364605E+02	5.6540	1.379087E+03	-1.278529E+01	1.385524E+03
7	-5.000000E-01	1.812767E+03	-1.769386E+02	-9.809570E+01	-2.8157	1.817592E+03	-1.817632E+02	1.914954E+03
	5.000000E-01	1.812767E+03	-1.769386E+02	-9.809570E+01	-2.8157	1.817592E+03	-1.817632E+02	1.914954E+03
6	-5.000000E-01	1.349300E+03	3.807900E+02	1.146431E+01	0.6781	1.349436E+03	3.806543E+02	1.205075E+03
	5.000000E-01	1.349300E+03	3.807900E+02	1.146431E+01	0.6781	1.349436E+03	3.806543E+02	1.205075E+03
203 CEN/8	-5.000000E-01	1.399866E+03	8.974188E+01	-8.790172E+01	-3.8214	1.405737E+03	8.387050E+01	1.365735E+03
	5.000000E-01	1.399866E+03	8.974188E+01	-8.790172E+01	-3.8214	1.405737E+03	8.387050E+01	1.365735E+03
6	-5.000000E-01	1.349300E+03	3.807900E+02	-1.146431E+01	-0.6781	1.349436E+03	3.806543E+02	1.205075E+03
	5.000000E-01	1.349300E+03	3.807900E+02	-1.146431E+01	-0.6781	1.349436E+03	3.806543E+02	1.205075E+03
7	-5.000000E-01	1.812767E+03	-1.769386E+02	9.809570E+01	2.8157	1.817592E+03	-1.817632E+02	1.914954E+03
	5.000000E-01	1.812767E+03	-1.769386E+02	9.809570E+01	2.8157	1.817592E+03	-1.817632E+02	1.914954E+03
11	-5.000000E-01	1.365577E+03	7.245544E-01	-1.364605E+02	-5.6540	1.379087E+03	-1.278529E+01	1.385524E+03
	5.000000E-01	1.365577E+03	7.245544E-01	-1.364605E+02	-5.6540	1.379087E+03	-1.278529E+01	1.385524E+03
9	-5.000000E-01	1.071818E+03	1.543916E+02	-3.017778E+02	-16.6700	1.162184E+03	6.402594E+01	1.131531E+03
	5.000000E-01	1.071818E+03	1.543916E+02	-3.017778E+02	-16.6700	1.162184E+03	6.402594E+01	1.131531E+03

Exhibit 12.6 (continued)   Stresses in quadrilateral elements (QUAD8)

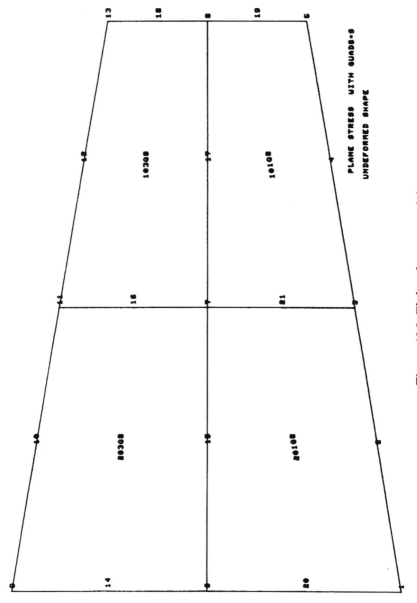

Figure 12.3 Finite element model.

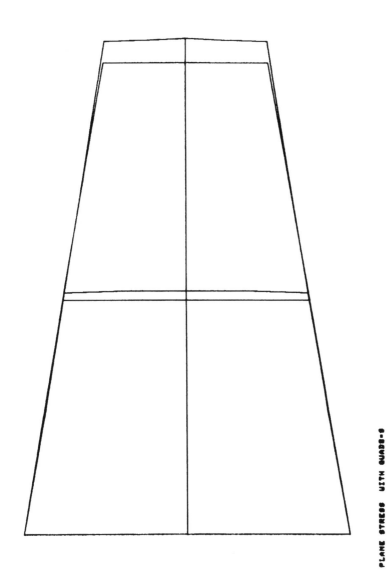

PLANE STRESS WITH QUAD8-8
STATIC DEFOR. SUBCASE 1   LOAD 11

Figure 12.4 Deformed shape of the thin sheet.

13
Problem 13

13.1 Statement of the problem

Consider the solid block described in Figure 13.1. It is made of concrete and it is loaded with a uniform pressure of 100 Kilos per square meter (K/m^2) on the top surface. Determine the stresses in the block and the foundation reactions due to this load. Assume $E=2.5\times10^9$ K/m^2 and $\nu=0.25$ for concrete.

13.2 Cards introduced

<u>Case Control Deck</u>   None

<u>Bulk Data Deck</u>   CHEXA
                                         PLOAD4
                                         PSOLID

13.3 MSC/NASTRAN formulation

The block is modeled using HEXA elements. The HEXA is a solid element with eight corner GRIDs and six faces.

Since the structure is symmetric only a half of it needs to be modeled and appropriate boundary conditions (displacement in the x-direction equals 0) are applied for all GRIDs on the symmetry plane.

Solid elements in MSC/NASTRAN do not support rotations; only three degrees of freedom per GRID are allowed. Therefore, degrees of freedom 4, 5, and 6 must be constrained for all solid elements.

13.4 Input Data Deck

The Executive Control Deck is shown in Exhibit 13.1. The Case Control Deck appears in Exhibit 13.2. The first part of it should be familiar at this point. The SPC=1111 card points to SPC1 cards in the Bulk Data Deck to enforce the boundary conditions and to suppress the rotations at all GRIDs. ECHO=UNSORT causes the Bulk Data Deck to be printed in the output file only once, just the way it was entered in the input file. (Remember that ECHO=BOTH causes the Bulk Data Deck to also be printed sorted by alphabetical order.) The stresses are to be printed for only one element --element 1 at the bottom of the block-- to avoid voluminous output.

The VIEW statement in the plot subdeck, without specifying explicitly α, β, or γ (see Figure 1.3), selects the default values (0.0°, 23.17°, 34.27°). The PLOT card produces a plot of the model without showing the element IDs or the grid IDs. PLOT LABEL BOTH creates the same plot including the labels. Examine the plots at the end of this chapter (Figures 13.2 and 13.3) to get an idea of what the model looks like. This will facilitate the discussion.

The Bulk Data Deck is presented in Exhibit 13.3. First examine the GRID cards. Observe that no constraints are specified directly on the GRID cards. SPC1 cards are used for this purpose.

The solid elements are defined using CHEXA cards. Each card gives the ID number of the eight GRIDs that define the HEXA, and the ID number of the MAT1 card on which the material characteristics of the HEXA element are supplied. Notice that GRIDs 9 to 20 are optional. In this example mid-side nodes are not included; therefore, only eight GRIDs per element are required.

Each CHEXA card actually consists of two "physical" cards. Thus, the tenth field of the first card must indicate that there is a second (continuation) card. This is done simply by repeating in the first field of the second card the same entry as the tenth field of the first card. (Remember that what follows the plus sign is arbitrary).

The boundary conditions are determined using SPC1 cards. The first SPC1 card is used to constrain all rotations for all GRID points. It is assumed that the block is prevented from moving in any direction at the base; therefore, all translational degrees of freedom are suppressed at the bottom of the block. This is accomplished with the second SPC1 card. The third group of SPC1 cards is used to apply the symmetry condition. The displacement in the x-direction is constrained for all grids on the symmetry plane (y-z plane).

The PSOLID card gives some additional information regarding the solid elements. The third field supplies the ID number (333333) of the material property card. The fourth field is used to specify the so-called material coordinate system. Since this is a homogeneous and isotropic block (same properties in all directions) there is no need to define a special material coordinate system. For simplicity, a 0 (not a blank!) is entered here to select the basic coordinate system as the material coordinate system. The fifth field contains the number 3. This means that a 3×3×3 (=27) Gaussian integration scheme is employed to determine the HEXA stiffness matrix.

The PLOAD4 cards, selected by LOAD=44 in the Case Control Deck, define the pressure load on top of the block. Examine the first PLOAD4 card. The 25 in the third field states that the pressure is acting on one of the faces of the HEXA whose ID is 25. The entry 100. in the fourth field specifies that the pressure is 100 K/m². Since the pressure on the face is constant, P2, P3, and P4 need not to be supplied. GRID IDs 500 and 506 identify the face of the HEXA on which the pressure is applied, by referencing one diagonal of the face. The other PLOAD4 cards can be interpreted similarly.

13. 5 Results

Exhibit 13.4 shows the value of EPSILON. Exhibit 13.5 shows the reactions at the supports. It also shows the forces in the x-direction at the GRIDs located on the symmetry plane (GRIDs 100, 105, 110, 200, 205, 210, 300, 305, 310, 400, 405, 410, 500, 505, and 510).

Exhibit 13.6 displays the value of the components of the stress tensor (σ_{xx}, σ_{yy}, σ_{zz}, σ_{xy}, σ_{xz}, σ_{yz}) for one HEXA (in this case HEXA number 1). Stresses are given at the center of the element and at the corner nodes. These stresses are given in the so-called material coordinate system, which in this case is the basic coordinate system as a result of the choice made on the PSOLID card (0 in field four). The last entry shows the Von Mises stress (σ_{VM}), defined in terms of the principal stresses as

$$\sigma_{VM} = \{1/2[(\sigma_1 - \sigma_2)^2 + (\sigma_2 - \sigma_3)^2 + (\sigma_3 - \sigma_1)^2]\}^{1/2} \qquad (13.1)$$

The column labeled mean pressure gives the value p, defined as

$$p = -(\sigma_1 + \sigma_2 + \sigma_3)/3 \qquad (13.2)$$

The plots showing the model geometry are presented in Figures 13.2 and 13.3.

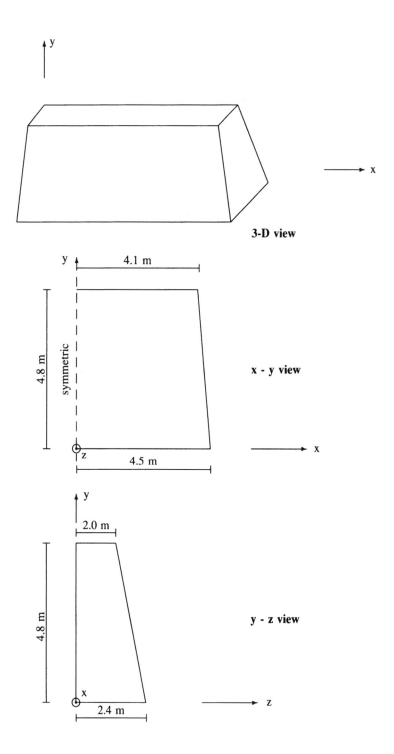

3-D view

x - y view

y - z view

Figure 13.1

```
N A S T R A N   E X E C U T I V E   C O N T R O L   D E C K   E C H O


ID    SOLID,SOLID
TIME    5
SOL 24
CEND
```

Exhibit 13.1   Executive Control Deck

```
              C A S E    C O N T R O L   D E C K   E C H O
CARD
COUNT
  1       TITLE=WALL SUBJECTED TO UNIFORM LOAD USING SYMMETRIC BOUN. CON.
  2       SET 22= 1
  3       STRESS=22
  4       SPCFORCES= ALL
  5       LOAD =44
  6       SPC=1111
  7       ECHO=UNSORT
  8       $$
  9       $$  P L O T S
 10       $$
 11       OUTPUT(PLOT)
 12       SET  11  INCLUDE  ALL
 13       AXES   Z,  X,  Y
 14       VIEW
 15       FIND
 16       PLOT
 17       PLOT    LABEL  BOTH
 18       $$
 19       BEGIN    BULK
```

Exhibit 13.2   Case Control Deck

	I N P U T	B U L K	D A T A	D E C K	E C H O				
. 1 ..	2 ..	3 ..	4 ..	5 ..	6 ..7 ..8 .. 9 .. 10				
$$									
GRID	100		0.0	0.0	0.0				
GRID	101		1.0	0.0	0.0				
GRID	102		2.0	0.0	0.0				
GRID	103		3.0	0.0	0.0				
GRID	104		4.5	0.0	0.0				
GRID	105		0.0	0.0	1.0				
GRID	106		1.0	0.0	1.0				
GRID	107		2.0	0.0	1.0				
GRID	108		3.0	0.0	1.0				
GRID	109		4.5	0.0	1.0				
GRID	110		0.0	0.0	2.4				
GRID	111		1.0	0.0	2.4				
GRID	112		2.0	0.0	2.4				
GRID	113		3.0	0.0	2.4				
GRID	114		4.5	0.0	2.4				
$$									
GRID	200		0.0	1.2	0.0				
GRID	201		1.0	1.2	0.0				
GRID	202		2.0	1.2	0.0				
GRID	203		3.0	1.2	0.0				
GRID	204		4.4	1.2	0.0				
GRID	205		0.0	1.2	1.0				
GRID	206		1.0	1.2	1.0				
GRID	207		2.0	1.2	1.0				
GRID	208		3.0	1.2	1.0				
GRID	209		4.4	1.2	1.0				
GRID	210		0.0	1.2	2.3				
GRID	211		1.0	1.2	2.3				
GRID	212		2.0	1.2	2.3				
GRID	213		3.0	1.2	2.3				
GRID	214		4.4	1.2	2.3				

Exhibit 13.3   Input Bulk Data Deck

```
              I N P U T   B U L K   D A T A   D E C K   E C H O
.   1 ..   2  ..   3  ..   4  ..   5  ..   6  .. 7  .. 8  ..  9  ..  10
$$
GRID      300           0.0    2.4    0.0
GRID      301           1.0    2.4    0.0
GRID      302           2.0    2.4    0.0
GRID      303           3.0    2.4    0.0
GRID      304           4.3    2.4    0.0
GRID      305           0.0    2.4    1.0
GRID      306           1.0    2.4    1.0
GRID      307           2.0    2.4    1.0
GRID      308           3.0    2.4    1.0
GRID      309           4.3    2.4    1.0
GRID      310           0.0    2.4    2.2
GRID      311           1.0    2.4    2.2
GRID      312           2.0    2.4    2.2
GRID      313           3.0    2.4    2.2
GRID      314           4.3    2.4    2.2
$$
GRID      400           0.0    3.6    0.0
GRID      401           1.0    3.6    0.0
GRID      402           2.0    3.6    0.0
GRID      403           3.0    3.6    0.0
GRID      404           4.2    3.6    0.0
GRID      405           0.0    3.6    1.0
GRID      406           1.0    3.6    1.0
GRID      407           2.0    3.6    1.0
GRID      408           3.0    3.6    1.0
GRID      409           4.2    3.6    1.0
GRID      410           0.0    3.6    2.1
GRID      411           1.0    3.6    2.1
GRID      412           2.0    3.6    2.1
GRID      413           3.0    3.6    2.1
GRID      414           4.2    3.6    2.1
```

Exhibit 13.3 (continued)   Input Bulk Data Deck

```
              I N P U T   B U L K   D A T A   D E C K   E C H O
   .  1 ..  2 ..  3 ..  4 ..  5 ..  6  .. 7 .. 8 ..  9  .. 10
$$
GRID      500            0.0     4.8     0.0
GRID      501            1.0     4.8     0.0
GRID      502            2.0     4.8     0.0
GRID      503            3.0     4.8     0.0
GRID      504            4.1     4.8     0.0
GRID      505            0.0     4.8     1.0
GRID      506            1.0     4.8     1.0
GRID      507            2.0     4.8     1.0
GRID      508            3.0     4.8     1.0
GRID      509            4.1     4.8     1.0
GRID      510            0.0     4.8     2.0
GRID      511            1.0     4.8     2.0
GRID      512            2.0     4.8     2.0
GRID      513            3.0     4.8     2.0
GRID      514            4.1     4.8     2.0
$$
$$$$$$$$$$$$$$$$$$$$  S O L I D   E L E M E N T S
$$
CHEXA      1     333    100     101     106    105   200   201+101
+101      206    205
CHEXA      2     333    101     102     107    106   201   202+102
+102      207    206
CHEXA      3     333    102     103     108    107   202   203+103
+103      208    207
CHEXA      4     333    103     104     109    108   203   204+104
+104      209    208
CHEXA      5     333    105     106     111    110   205   206+105
+105      211    210
CHEXA      6     333    106     107     112    111   206   207+106
+106      212    211
CHEXA      7     333    107     108     113    112   207   208+107
+107      213    212
CHEXA      8     333    108     109     114    113   208   209+108
+108      214    213
CHEXA      9     333    200     201     206    205   300   301+109
```

Exhibit 13.3 (continued)   Input Bulk Data Deck

	I N P U T	B U L K	D A T A	D E C K	E C H O				
. 1 ..	2 ..	3 ..	4 ..	5 ..	6 ..	7 ..	8 ..	9 ..	10
+109	306	305							
CHEXA	10	333	201	202	207	206	301	302+110	
+110	307	306							
CHEXA	11	333	202	203	208	207	302	303+111	
+111	308	307							
CHEXA	12	333	203	204	209	208	303	304+112	
+112	309	308							
CHEXA	13	333	205	206	211	210	305	306+113	
+113	311	310							
CHEXA	14	333	206	207	212	211	306	307+114	
+114	312	311							
CHEXA	15	333	207	208	213	212	307	308+115	
+115	313	312							
CHEXA	16	333	208	209	214	213	308	309+116	
+116	314	313							
CHEXA	17	333	300	301	306	305	400	401+117	
+117	406	405							
CHEXA	18	333	301	302	307	306	401	402+118	
+118	407	406							
CHEXA	19	333	302	303	308	307	402	403+119	
+119	408	407							
CHEXA	20	333	303	304	309	308	403	404+120	
+120	409	408							
CHEXA	21	333	305	306	311	310	405	406+121	
+121	411	410							
CHEXA	22	333	306	307	312	311	406	407+122	
+122	412	411							
CHEXA	23	333	307	308	313	312	407	408+123	
+123	413	412							
CHEXA	24	333	308	309	314	313	408	409+124	
+124	414	413							
CHEXA	25	333	400	401	406	405	500	501+125	
+125	506	505							
CHEXA	26	333	401	402	407	406	501	502+126	
+126	507	506							
CHEXA	27	333	402	403	408	407	502	503+127	
+127	508	507							
CHEXA	28	333	403	404	409	408	503	504+128	
+128	509	508							
CHEXA	29	333	405	406	411	410	505	506+129	
+129	511	510							

Exhibit 13.3 (continued)   Input Bulk Data Deck

```
                  I N P U T   B U L K   D A T A   D E C K   E C H O
  .   1 ..    2 ..   3 ..   4 ..   5 ..   6 ..   7 ..   8 ..   9 ..  10
CHEXA          30    333    406    407    412    411    506    507+130
+130          512    511
CHEXA          31    333    407    408    413    412    507    508+131
+131          513    512
CHEXA          32    333    408    409    414    413    508    509+132
+132          514    513
$$
$$          BOUNDARY CONDITIONS
$$
$$          SOLID ELEMENTS DO NOT SUPPORT ROTATIONS
$$
SPC1   1111    456    100    THRU    514
$$
$$          DISPLACEMENTS ARE CONSTRAINED AT THE FOUNDATION
$$
SPC1   1111    123    100    THRU    114
$$
$$          SYMMETRIC BOUNDARY CONDITION
$$
SPC1   1111      1    100    200    300    400    500
SPC1   1111      1    105    205    305    405    505
SPC1   1111      1    110    210    310    410    510
$$
$$
$$
PSOLID,333,333333,0  ,3,
MAT1,333333,2.5+9,   ,.25
$$$$$$$
$$$$$$$   LOADS
$$$$$$$
PLOAD4  44     25    100.                          500    506
PLOAD4  44     29    100.                          505    511
PLOAD4  44     26    100.                          501    507
PLOAD4  44     30    100.                          511    507
PLOAD4  44     27    100.                          502    508
PLOAD4  44     31    100.                          512    508
PLOAD4  44     28    100.                          504    508
PLOAD4  44     32    100.                          509    513
ENDDATA
```

Exhibit 13.3 (continued)   Input Bulk Data Deck

*** USER INFORMATION MESSAGE 5293 FOR DATA BLOCK KLL

LOAD SEQ. NO.	EPSILON	EXTERNAL WORK	EPSILONS LARGER THAN 0.001 ARE FLAGGED WITH ASTERISKS
1	1.0028715E-16	6.8134519E-05	

Exhibit 13.4 Epsilon

F O R C E S   O F   S I N G L E - P O I N T   C O N S T R A I N T

POINT ID.	TYPE	T1	T2	T3	R1	R2	R3
100	G	8.476781E+00	2.782708E+01	8.301102E+00	0.0	0.0	0.0
101	G	-1.967141E+00	5.528505E+01	1.650519E+01	0.0	0.0	0.0
102	G	-4.203514E+00	5.403203E+01	1.615696E+01	0.0	0.0	0.0
103	G	-8.526905E+00	6.517274E+01	1.919923E+01	0.0	0.0	0.0
104	G	-1.344799E+01	4.242160E+01	1.109027E+01	0.0	0.0	0.0
105	G	1.509919E+01	4.431424E+01	-1.699141E+00	0.0	0.0	0.0
106	G	-4.137254E+00	8.817751E+01	-3.380742E+00	0.0	0.0	0.0
107	G	-8.697078E+00	8.666246E+01	-3.356628E+00	0.0	0.0	0.0
108	G	-1.718519E+01	1.073700E+02	-4.835208E+00	0.0	0.0	0.0
109	G	-2.622451E+01	7.392403E+01	-4.008934E+00	0.0	0.0	0.0
110	G	6.102201E+00	1.880913E+01	-5.875799E+00	0.0	0.0	0.0
111	G	-2.014733E+00	3.749233E+01	-1.171291E+01	0.0	0.0	0.0
112	G	-4.143954E+00	3.710635E+01	-1.159552E+01	0.0	0.0	0.0
113	G	-7.785521E+00	4.732445E+01	-1.489747E+01	0.0	0.0	0.0
114	G	-1.176084E+01	3.408095E+01	-9.890396E+00	0.0	0.0	0.0

Exhibit 13.5 Single point constraint forces

F O R C E S   O F   S I N G L E - P O I N T   C O N S T R A I N T

POINT ID.	TYPE	T1	T2	T3	R1	R2	R3
200	G	1.033193E+01	0.0	0.0	0.0	0.0	0.0
205	G	1.823698E+01	0.0	0.0	0.0	0.0	0.0
210	G	6.472505E+00	0.0	0.0	0.0	0.0	0.0
300	G	5.882340E+00	0.0	0.0	0.0	0.0	0.0
305	G	1.057977E+01	0.0	0.0	0.0	0.0	0.0
310	G	3.531907E+00	0.0	0.0	0.0	0.0	0.0
400	G	4.045879E+00	0.0	0.0	0.0	0.0	0.0
405	G	8.411951E+00	0.0	0.0	0.0	0.0	0.0
410	G	4.030748E+00	0.0	0.0	0.0	0.0	0.0
500	G	1.935836E+00	0.0	0.0	0.0	0.0	0.0
505	G	4.491709E+00	0.0	0.0	0.0	0.0	0.0
510	G	2.464886E+00	0.0	0.0	0.0	0.0	0.0

Exhibit 13.5 (continued)   Single point constraint forces

S T R E S S E S   I N   H E X A H E D R O N   S O L I D   E L E M E N T S   (H E X A)

ELEMENT-ID	CORNER GRID-ID	-----CENTER AND CORNER POINT STRESSES-----			DIR. COSINES -A- -B- -C-	MEAN PRESSURE	VON MISES
		NORMAL	SHEAR	PRINCIPAL			
1	OGRID CS	8 GP					
	CENTER	X -2.336415E+01	XY 1.750558E+00	A -1.447024E+01	LX-0.02-0.02-1.00	4.556467E+01	8.036463E+01
		Y -9.851999E+01	YZ -5.316312E+00	B -9.889650E+01	LY-0.06 1.00-0.02		
		Z -1.480987E+01	ZX -6.166665E-02	C -2.332726E+01	LZ 1.00 0.06-0.02		

Exhibit 13.6   Stresses in hexahedron solid elements (HEXA)

ELEMENT-ID	CORNER GRID-ID		------CENTER AND CORNER POINT STRESSES------- NORMAL		SHEAR		PRINCIPAL	DIR. COSINES -A- -B- -C-		MEAN PRESSURE	VON MISES
100	X	-4.141453E+01	XY	2.948582E-17	A	-3.787933E+01	LX 0.00 0.00-1.00		6.902422E+01	8.818503E+01
	Y	-1.242436E+02	YZ	-1.747328E+01	B	-1.277788E+02	LY-0.20 0.98 0.00			
	Z	-4.141453E+01	ZX	-7.185744E-19	C	-4.141453E+01	LZ 0.98 0.20 0.00			
101	X	-4.121125E+01	XY	3.513237E+00	A	-3.757649E+01	LX-0.19-0.04-0.98		6.868542E+01	8.793102E+01
	Y	-1.236338E+02	YZ	-1.733364E+01	B	-1.272685E+02	LY-0.20 0.98 0.00			
	Z	-4.121125E+01	ZX	1.314608E-18	C	-4.121125E+01	LZ 0.96 0.20-0.20			
106	X	-2.981251E+01	XY	3.111181E+00	A	-2.885899E+01	LX-0.41-0.05-0.91		4.968752E+01	6.106090E+01
	Y	-8.943754E+01	YZ	-6.934248E+00	B	-9.039106E+01	LY-0.12 0.99 0.00			
	Z	-2.981251E+01	ZX	5.564713E-19	C	-2.981251E+01	LZ 0.91 0.11-0.41			
105	X	-2.992408E+01	XY	-5.445310E-18	A	-2.911827E+01	LX 0.00 0.00-1.00		4.987347E+01	6.106086E+01
	Y	-8.977224E+01	YZ	-6.991113E+00	B	-9.057805E+01	LY-0.11 0.99 0.00			
	Z	-2.992408E+01	ZX	-1.275099E-17	C	-2.992408E+01	LZ 0.99 0.11 0.00			
200	X	-1.618828E+01	XY	2.439333E-01	A	6.642361E-01	LX 0.01 0.00-1.00		4.103341E+01	1.008744E+02
	Y	-1.074491E+02	YZ	-3.684735E+00	B	-1.075754E+02	LY-0.03 1.00 0.00			
	Z	5.371475E-01	ZX	1.675634E-01	C	-1.618910E+01	LZ 1.00 0.03 0.01			
201	X	-1.608433E+01	XY	3.757170E+00	A	5.785592E-01	LX-0.03-0.04-1.00		4.086016E+01	1.005579E+02
	Y	-1.069386E+02	YZ	-3.655152E+00	B	-1.072167E+02	LY-0.03 1.00-0.04			
	Z	4.424473E-01	ZX	-3.149040E-01	C	-1.594232E+01	LZ 1.00 0.03-0.03			
206	X	-6.132989E+00	XY	3.245061E+00	A	1.189446E+01	LX-0.01-0.05-1.00		2.266637E+01	7.841804E+01
	Y	-7.322485E+01	YZ	6.744241E+00	B	-7.391674E+01	LY 0.08 1.00-0.05			
	Z	1.135872E+01	ZX	-4.142300E-01	C	-5.976836E+00	LZ 1.00-0.08 0.00			
205	X	-6.145231E+00	XY	1.338809E-01	A	1.208559E+01	LX 0.00 0.00-1.00		2.268677E+01	7.857332E+01
	Y	-7.346022E+01	YZ	6.797429E+00	B	-7.400059E+01	LY 0.08 1.00 0.00			
	Z	1.154513E+01	ZX	6.823741E-02	C	-6.145329E+00	LZ 1.00-0.08 0.00			

Exhibit 13.6 (continued)   Stresses in hexahedron solid elements (HEXA)

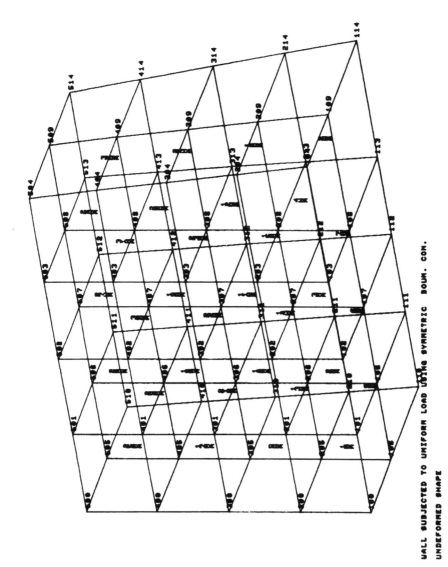

WALL SUBJECTED TO UNIFORM LOAD USING SYMMETRIC BOUN. CON.
UNDEFORMED SHAPE

Figure 13.2 Finite element model of the solid block (labelled).

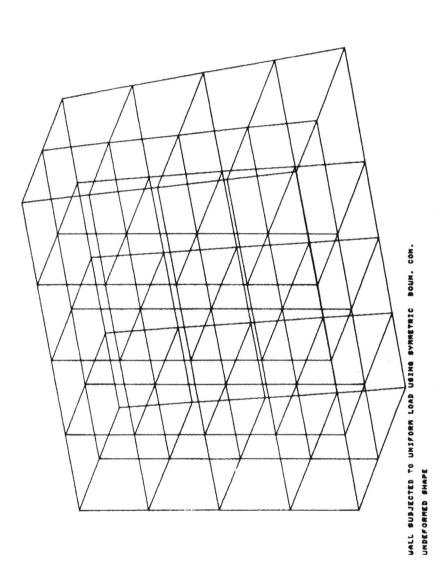

WALL SUBJECTED TO UNIFORM LOAD USING SYMMETRIC BOUN. CON.
UNDEFORMED SHAPE

Figure 13.3 Finite element model of the solid block (unlabelled).

PART 2

DYNAMICS

14

Problem 14

14.1 Statement of the problem

Consider the beam shown in Figure 14.1. It is mounted on two springs, each with a spring constant $k = 10^4$ lb/in, and it has a concentrated mass of 0.02 lb-s^2/in located at five inches from one end. The material properties are as follows: $E = 30 \times 10^6$ psi; $\nu = 0.29$; and $\rho = .0007$ lb-s^2/in^4. Determine the normal modes and natural frequencies of this structure.

14.2 Cards introduced

Case Control Deck   METHOD

Bulk Data Deck   CMASS2
                                           EIGR

14.3 MSC/NASTRAN formulation

The beam is modeled with eight BAR elements as shown in Figure 14.2. The springs are defined using a CELAS2 card.

Determination of the normal modes and natural frequencies of a structure requires the solution of the following equation

$$[M]d^2\{x\}/dt^2 + [K]\{x\} = \{0\} \qquad (14.1)$$

which reduces to solving a linear eigenvalue problem of the form

$$([K]-\lambda[M])\{\phi\} = \{0\} \qquad (14.2)$$

where [M] is the mass matrix of the structure, [K] is the stiffness matrix, $\{x\}$ is the displacement vector ($\{x\} = \{\phi\}\cos(\omega t)$), $\{\phi\}$ is the eigenvector or normal mode, and λ is the corresponding eigenvalue. Recall that a natural frequency (ω) is the square root of its corresponding eigenvalue.

This type of analysis is performed with MSC/NASTRAN using SOL 3. The inverse power method will be used to compute the eigenvalues.

14.4 Input Data Deck

The Executive Control Deck, shown in Exhibit 14.1, selects SOL 3 (normal modes). Examine the Case Control Deck shown in Exhibit 14.2. The card DISP = ALL has a slightly different meaning in this context (remember that in SOL 24 it was used to request the displacements to be printed). In running a normal modes analysis, it requests the mode shapes to be printed.

The normal modes, even though they are expressed as vectors having components of displacement in six directions, are not displacements in the strict sense of the word. They are just patterns of deformation ("shapes") that show the relative displacements at different locations of the structure as it vibrates freely.

The METHOD card is always needed in the Case Control Deck when performing a normal modes analysis. It selects an EIGR card in the Bulk Data Deck on which certain parameters of the eigenvalue extraction technique are specified.

The statements to make the plots are similar to those demonstrated in SOL 24 runs. There is only one new instruction: PLOT MODAL DEFORMATION 0, 1 THRU 5.[1] This statement requests the first five normal modes to be plotted.

Exhibits 14.3 and 14.4 show the Bulk Data Deck. The GRID, CBAR, PBAR, and MAT1 cards have already been discussed in previous examples. Notice that to generate the mass matrix of the system, the MAT1 card must specify ρ (mass per unit of volume) in field six .

Only two degrees of freedom per GRID (displacement in the y-direction and rotations about z) are needed in order to properly represent the way this beam deforms. The moment of inertia (according to the convention explained in Problem 7, Figure 7.3) associated with this type of behavior is I_2. Thus, I_1 need not be supplied. Consequently, the entry corresponding to I_1 is blank on the PBAR card, and only degrees of freedom 2 and 6 are left free on the GRID cards.

The CMASS2 card is used to specify the concentrated mass. This card defines a mass of .02 lb-s^2/in attached to GRID 102. The number 2 in field 5

---

[1] This 0 is to display the undeformed structure in the plot. If omitted, only the mode shape is plotted.

refers to degree of freedom 2 (y-displacement). This means that the mass resists acceleration in the y-direction. In a three dimensional problem there probably would be a mass resisting acceleration in all three directions x, y, and z.

The springs are defined with CELAS2 cards. The value of the spring constant goes in the third field of the CELAS2 card, the GRID point to which the spring is connected goes in field 4, and the degree of freedom associated with the spring goes in field 5.

Finally, examine the EIGR card, which is required in all SOL 3 runs. This card is selected in the Case Control Deck by a METHOD card (METHOD=99 in this example). The entry in the third field of the EIGR card chooses the method used for determining the eigenvalues of the problem. SINV corresponds to the inverse power method. F_1 and F_2 (fields 4 and 5) determine the range of interest for the frequencies. F_1 and F_2 are expressed in cycles/second (Hz). When choosing SINV, field 6 of this card is not used. Field 7 is blank, indicating that all the frequencies in the (F_1, F_2) range are to be computed. This is the safest approach since in general the number of frequencies in the (F_1, F_2) interval is not known in advanced.

14.5 Results

The output file does not show the EPSILON that characterized all SOL 24 runs. EPSILON is just a numerical conditioning indicator used for static analysis that has no counterpart in SOL 3 -- or in any other type of analysis.

Exhibit 14.5 shows a message indicating that all the eigenvalues in the range specified were computed. Exhibit 14.6 shows the natural frequencies. Note that two extra frequencies outside the range of interest (0 to 800 Hz) were obtained (1,313.166 Hz and 1,965.976 Hz). This happens very often with iterative techniques like the inverse power method: when the initial point for the iteration process is selected (this is done automatically by the program) it is not possible to know to which value it is going to converge.

Notice that only 1's appear under the GENERALIZED MASS heading, because the eigenvectors (normal modes) have been scaled in such a way that

$$\{\phi\}^T[M]\{\phi\} = 1 \qquad (14.3)$$

This is the default option for scaling the eigenvectors. The column labeled GENERALIZED STIFFNESS shows, for each normal mode (eigenvector), the product

$$\{\phi\}^T[K]\{\phi\} = \omega^2 = \lambda \qquad (14.4)$$

The eigenvectors are presented in Exhibit 14.7. The first plot (Figure 14.3), generated by PLOT LABEL BOTH, shows the model geometry. The remaining five plots (Figures 14.4 through 14.8) correspond to the first five normal modes.

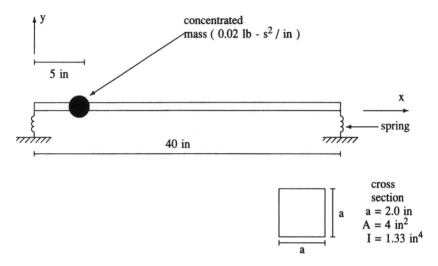

Figure 14.1

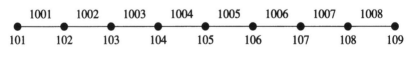

Figure 14.2

```
N A S T R A N   E X E C U T I V E   C O N T R O L   D E C K   E C H O


    ID   A,B
    TIME  5
    SOL   3        $  3  MEANS NORMAL MODES AND  NATURAL  FREQ.
    CEND
```

Exhibit 14.1   Executive Control Deck

```
                  C A S E    C O N T R O L   D E C K   E C H O
CARD
COUNT
  1      TITLE=              NORMAL  MODES AND  NAT. FREQ. OF A BEAM
  2      SUBTITLE=           MOUNTED ON SPRINGS
  3      $$
  4      DISP=ALL       $  PRINT  NORMAL  MODES
  5      $$
  6      ECHO=BOTH
  7      METHOD= 99     $  TO SELECT THE EIGR  CARD
  8      $$
  9      $$   PLOTS
 10      $$
 11      OUTPUT(PLOT)
 12      SET  55   INCLUDE   ALL
 13      AXES   Z,X,Y
 14      VIEW  0.0,0.0,0.0
 15      FIND
 16      PLOT   LABEL   BOTH
 17      PLOT   MODAL   DEFORMATION  0 ,1   THRU   5
 18      $$
 19      BEGIN BULK
```

Exhibit 14.2   Case Control Deck

```
                I N P U T   B U L K   D A T A   D E C K   E C H O
.   1  ..   2  ..   3  ..   4  ..   5  ..   6  ..   7  ..   8  ..   9
$$
GRID         101         00.00   0.00    0.00            1345
GRID         102         5.      0.00    0.00            1345
GRID         103         10.00   0.00    0.00            1345
GRID         104         15.     0.00    0.00            1345
GRID         105         20.00   0.00    0.00            1345
GRID         106         25.00   0.00    0.00            1345
GRID         107         30.00   0.00    0.00            1345
GRID         108         35.00   0.00    0.00            1345
GRID         109         40.     0.00    0.00            1345
$$
CBAR         1001 5555       101     102     1.000   0.000   1.000
CBAR         1002 5555       102     103     1.000   0.000   1.000
CBAR         1003 5555       103     104     1.000   0.000   1.000
CBAR         1004 5555       104     105     1.000   0.000   1.000
CBAR         1005 5555       105     106     1.000   0.000   1.000
CBAR         1006 5555       106     107     1.000   0.000   1.000
CBAR         1007 5555       107     108     1.000   0.000   1.000
CBAR         1008 5555       108     109     1.000   0.000   1.000
$$
PBAR    5555    8888    4.00            1.333
MAT1    8888    30.0+6          .29     7.0-4
$$
$$  CONCENTRATED MASS   AT POINT   102
CMASS2  123456  .02     102     2
$$
$$ SPRINGS AT NODE   101    AND   109
$$
CELAS2  2233    10.+4   101     2
CELAS2  2244    10.+4   109     2
$$
$$   SPECIFICATIONS FOR EIGENVALUE EXTRACTION
$$
EIGR    99      SINV    0.00    800.0
ENDDATA
```

Exhibit 14.3 Input Bulk Data Deck

```
                    S O R T E D   B U L K   D A T A   E C H O
CARD
COUNT   .   1  ..   2  ..   3  ..   4  ..   5  ..   6  ..   7  ..   8
  1-   CBAR    1001   5555    101    102   1.000  0.000  1.000
  2-   CBAR    1002   5555    102    103   1.000  0.000  1.000
  3-   CBAR    1003   5555    103    104   1.000  0.000  1.000
  4-   CBAR    1004   5555    104    105   1.000  0.000  1.000
  5-   CBAR    1005   5555    105    106   1.000  0.000  1.000
  6-   CBAR    1006   5555    106    107   1.000  0.000  1.000
  7-   CBAR    1007   5555    107    108   1.000  0.000  1.000
  8-   CBAR    1008   5555    108    109   1.000  0.000  1.000
  9-   CELAS2  2233   10.+4   101      2
 10-   CELAS2  2244   10.+4   109      2
 11-   CMASS2  123456 .02     102      2
 12-   EIGR    99     SINV    0.00   800.0
 13-   GRID    101            00.00  0.00   0.00               1345
 14-   GRID    102            5.     0.00   0.00               1345
 15-   GRID    103            10.00  0.00   0.00               1345
 16-   GRID    104            15.    0.00   0.00               1345
 17-   GRID    105            20.00  0.00   0.00               1345
 18-   GRID    106            25.00  0.00   0.00               1345
 19-   GRID    107            30.00  0.00   0.00               1345
 20-   GRID    108            35.00  0.00   0.00               1345
 21-   GRID    109            40.    0.00   0.00               1345
 22-   MAT1    8888   30.0+6          .29   7.0-4
 23-   PBAR    5555   8888    4.00          1.333
       ENDDATA
```

Exhibit 14.4   Bulk Data Deck sorted by alphabetical order

```
         EIGENVALUE   ANALYSIS   SUMMARY   (STURM INVERSE POWER)

    NUMBER OF EIGENVALUES EXTRACTED . . . . . .        6

    NUMBER OF TRIANGULAR DECOMPOSITIONS . . . .        4

    TOTAL NUMBER OF VECTOR ITERATIONS . . . . .       62

    REASON FOR TERMINATION:  ALL EIGENVALUES FOUND IN RANGE.
```

Exhibit 14.5   Eigenvalue analysis summary table

MODE NO.	EXTRACTION ORDER	EIGENVALUE	R E A L E I G E N V A L U E S RADIANS	CYCLES	GENERALIZED MASS	GENERALIZED STIFFNESS
1	1	4.011898E+05	6.333954E+02	1.008080E+02	1.000000E+00	4.011898E+05
2	2	2.870254E+06	1.694182E+03	2.696375E+02	1.000000E+00	2.870254E+06
3	3	8.406914E+06	2.899468E+03	4.614646E+02	1.000000E+00	8.406914E+06
4	4	2.469761E+07	4.969669E+03	7.909474E+02	1.000000E+00	2.469761E+07
5	5	6.807676E+07	8.250864E+03	1.313166E+03	1.000000E+00	6.807676E+07
6	6	1.525866E+08	1.235259E+04	1.965976E+03	1.000000E+00	1.525866E+08

Exhibit 14.6 Eigenvalues

EIGENVALUE = 4.011898E+05
CYCLES = 1.008080E+02   R E A L   E I G E N V E C T O R   N O.   1

POINT ID.	TYPE	T1	T2	T3	R1	R2	R3
101	G	0.0	7.474736E-01	0.0	0.0	0.0	2.532281E-01
102	G	0.0	1.975767E+00	0.0	0.0	0.0	2.305199E-01
103	G	0.0	2.991018E+00	0.0	0.0	0.0	1.708193E-01
104	G	0.0	3.645099E+00	0.0	0.0	0.0	8.780190E-02
105	G	0.0	3.849594E+00	0.0	0.0	0.0	-6.881803E-03
106	G	0.0	3.578767E+00	0.0	0.0	0.0	-1.000737E-01
107	G	0.0	2.873078E+00	0.0	0.0	0.0	-1.787324E-01
108	G	0.0	1.834544E+00	0.0	0.0	0.0	-2.315306E-01
109	G	0.0	6.146477E-01	0.0	0.0	0.0	-2.502036E-01

Exhibit 14.7 Eigenvectors

EIGENVALUE = 2.870254E+06
CYCLES = 2.696375E+02   REAL EIGENVECTOR NO. 2

POINT ID.	TYPE	T1	T2	T3	R1	R2	R3
101	G	0.0	-3.508466E+00	0.0	0.0	0.0	3.512619E-03
102	G	0.0	-3.344849E+00	0.0	0.0	0.0	9.114509E-02
103	G	0.0	-2.474962E+00	0.0	0.0	0.0	2.520099E-01
104	G	0.0	-9.245620E-01	0.0	0.0	0.0	3.529876E-01
105	G	0.0	8.839382E-01	0.0	0.0	0.0	3.513793E-01
106	G	0.0	2.417406E+00	0.0	0.0	0.0	2.466756E-01
107	G	0.0	3.247981E+00	0.0	0.0	0.0	8.034384E-02
108	G	0.0	3.236730E+00	0.0	0.0	0.0	-7.645603E-02
109	G	0.0	2.635057E+00	0.0	0.0	0.0	-1.422740E-01

EIGENVALUE = 8.406914E+06
CYCLES = 4.614646E+02   REAL EIGENVECTOR NO. 3

POINT ID.	TYPE	T1	T2	T3	R1	R2	R3
101	G	0.0	-4.349819E+00	0.0	0.0	0.0	4.432240E-01
102	G	0.0	-2.040447E+00	0.0	0.0	0.0	4.991749E-01
103	G	0.0	5.245880E-01	0.0	0.0	0.0	4.847203E-01
104	G	0.0	2.417733E+00	0.0	0.0	0.0	2.368521E-01
105	G	0.0	2.684499E+00	0.0	0.0	0.0	-1.361823E-01
106	G	0.0	1.160148E+00	0.0	0.0	0.0	-4.466740E-01
107	G	0.0	-1.442266E+00	0.0	0.0	0.0	-5.531805E-01
108	G	0.0	-4.048981E+00	0.0	0.0	0.0	-4.660812E-01
109	G	0.0	-6.117100E+00	0.0	0.0	0.0	-3.873952E-01

Exhibit 14.7 (continued) Eigenvectors

EIGENVALUE = 2.469761E+07
    CYCLES = 7.909474E+02

R E A L   E I G E N V E C T O R   N O .   4

POINT ID.	TYPE	T1	T2	T3	R1	R2	R3
101	G	0.0	5.540403E+00	0.0	0.0	0.0	-1.039036E+00
102	G	0.0	5.556005E-01	0.0	0.0	0.0	-9.128105E-01
103	G	0.0	-2.923901E+00	0.0	0.0	0.0	-3.883050E-01
104	G	0.0	-2.947293E+00	0.0	0.0	0.0	3.642945E-01
105	G	0.0	4.160675E-02	0.0	0.0	0.0	7.104319E-01
106	G	0.0	2.956192E+00	0.0	0.0	0.0	3.360690E-01
107	G	0.0	2.741470E+00	0.0	0.0	0.0	-4.347900E-01
108	G	0.0	-1.026209E+00	0.0	0.0	0.0	-9.863499E-01
109	G	0.0	-6.447567E+00	0.0	0.0	0.0	-1.133232E+00

EIGENVALUE = 6.807676E+07
    CYCLES = 1.313166E+03

R E A L   E I G E N V E C T O R   N O .   5

POINT ID.	TYPE	T1	T2	T3	R1	R2	R3
101	G	0.0	-6.487236E+00	0.0	0.0	0.0	1.788906E+00
102	G	0.0	1.184746E+00	0.0	0.0	0.0	1.025377E+00
103	G	0.0	2.650011E+00	0.0	0.0	0.0	-4.080639E-01
104	G	0.0	-1.268045E+00	0.0	0.0	0.0	-8.647950E-01
105	G	0.0	-3.683892E+00	0.0	0.0	0.0	6.690219E-02
106	G	0.0	-7.436818E-01	0.0	0.0	0.0	9.118040E-01
107	G	0.0	3.091102E+00	0.0	0.0	0.0	3.508782E-01
108	G	0.0	1.587568E+00	0.0	0.0	0.0	-9.165667E-01
109	G	0.0	-4.929048E+00	0.0	0.0	0.0	-1.496702E+00

Exhibit 14.7 (continued)   Eigenvectors

```
EIGENVALUE =  1.525866E+08
   CYCLES =  1.965976E+03          R E A L   E I G E N V E C T O R   N O .    6
POINT ID.  TYPE      T1              T2           T3          R1          R2          R3
   101      G       0.0        5.179650E+00      0.0         0.0         0.0      -1.961963E+00
   102      G       0.0       -2.017813E+00      0.0         0.0         0.0      -3.945526E-01
   103      G       0.0        1.005193E+00      0.0         0.0         0.0       1.035499E+00
   104      G       0.0        3.773324E+00      0.0         0.0         0.0      -2.727608E-01
   105      G       0.0       -9.682665E-01      0.0         0.0         0.0      -1.128531E+00
   106      G       0.0       -3.636172E+00      0.0         0.0         0.0       3.411893E-01
   107      G       0.0        1.195192E+00      0.0         0.0         0.0       1.061850E+00
   108      G       0.0        3.017398E+00      0.0         0.0         0.0      -5.964454E-01
   109      G       0.0       -4.046046E+00      0.0         0.0         0.0      -1.820810E+00
```

Exhibit 14.7 (continued) Eigenvectors

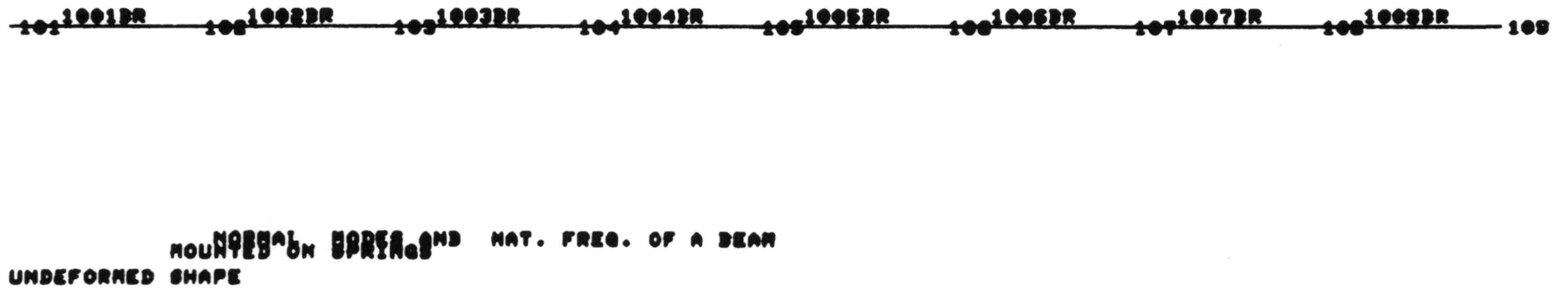

Figure 14.3 Finite element model of the beam.

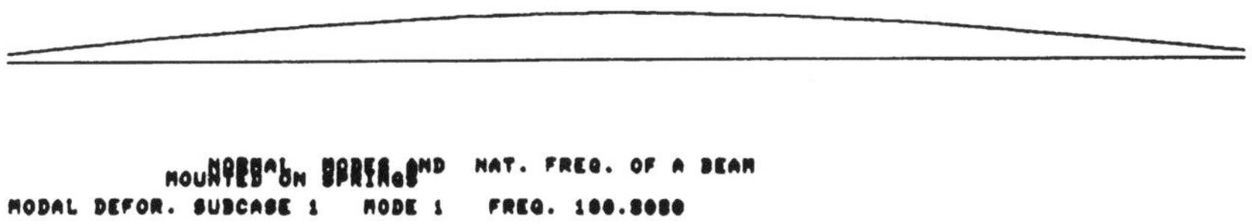
NORMAL MODES AND NAT. FREQ. OF A BEAM
MOUNTED ON SPRINGS
MODAL DEFOR. SUBCASE 1 MODE 1 FREQ.

Figure 14.4 First normal mode of the beam.

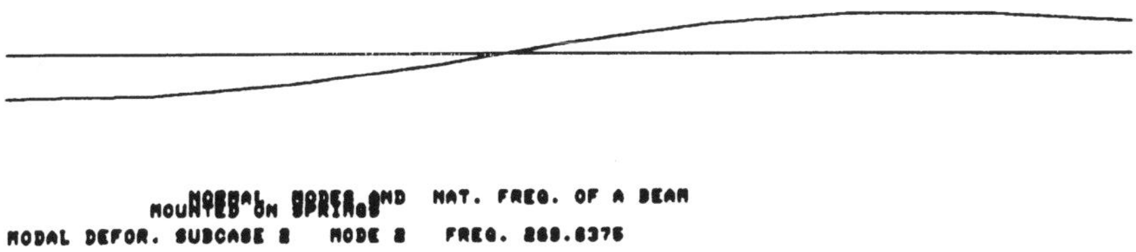
NORMAL MODES AND NAT. FREQ. OF A BEAM
MOUNTED ON SPRINGS
MODAL DEFOR. SUBCASE 2 MODE 2 FREQ.

Figure 14.5 Second normal mode of the beam.

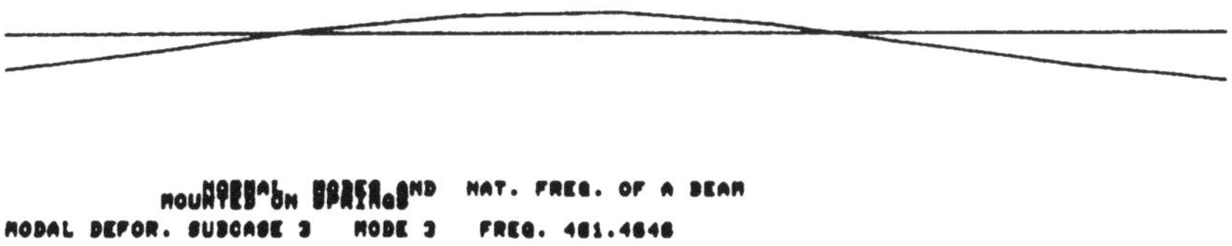
NORMAL MODES AND NAT. FREQ. OF A BEAM
MOUNTED ON SPRINGS
MODAL DEFOR. SUBCASE 3 MODE 3 FREQ.

Figure 14.6 Third normal mode of the beam.

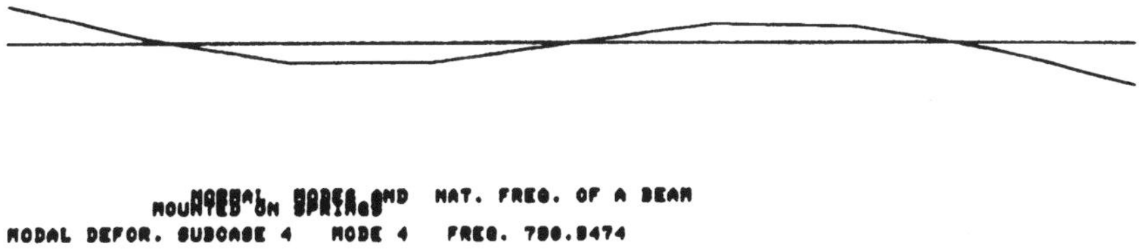
NORMAL MODES AND NAT. FREQ. OF A BEAM
MOUNTED ON SPRINGS
MODAL DEFOR. SUBCASE 4 MODE 4 FREQ.

Figure 14.7 Fourth normal mode of the beam.

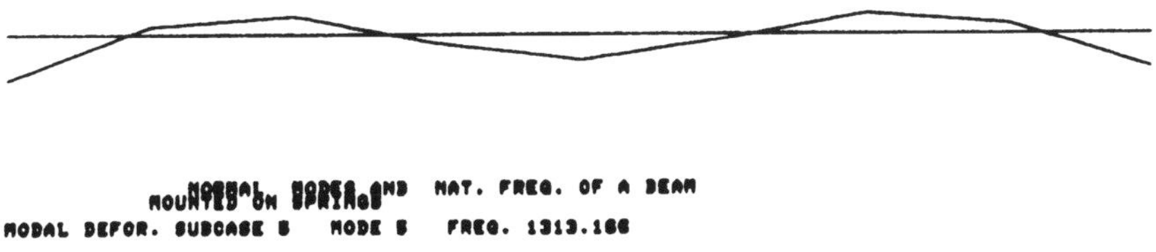
NORMAL MODES AND NAT. FREQ. OF A BEAM
MOUNTED ON SPRINGS
MODAL DEFOR. SUBCASE 5 MODE 5 FREQ. 1313.166

Figure 14.8 Fifth normal mode of the beam.

15

Problem 15

15.1 Statement of the problem

Consider the same structure analyzed in Problem 14 (Figure 14.1). Determine the normal modes and natural frequencies using the modified Givens method.

15.2 Cards introduced

Case Control Deck   None

Bulk Data Deck   None

15.3 MSC/NASTRAN formulation

The same model described in Problem 14 (Figure 14.2) is used. The modified Givens method, as opposed to the inverse power method employed in the previous example, is used to compute the natural frequencies. The modified Givens method, which is selected in MSC/NASTRAN by invoking MGIV on the EIGR card, is actually a combination of the Givens method (modified to treat the case of a singular mass matrix) and the QR technique for eigenvalue extraction.

In this example, as in the previous one, a lumped mass matrix is employed since no special requests are made in the Bulk Data Deck. The lumped mass matrix, which is the default option, is a diagonal matrix with non-zero terms associated only with translational degrees of freedom, i.e. no inertial effects associated with rotational degrees of freedom are considered.

15.4 Input Data Deck

The Executive Control Deck (Exhibit 15.1) selects SOL 3, which corresponds to normal modes analysis regardless of the method employed to calculate the natural frequencies.

The Case Control Deck (Exhibit 15.2) is similar to that of Problem 14 except for the plot statements that were deleted.

Notice a difference in the Bulk Data Deck (Exhibits 15.3 and 15.4) with respect to Problem 14 on the EIGR card. The modified Givens technique (MGIV) is selected in the third field. This method, in contrast to SINV (an iterative technique) determines all the eigenvalues, and therefore all the natural frequencies, simultaneously. Thus, there is no need to specify the range of interest for the frequencies. It is necessary to specify the number of eigenvectors (normal modes) to be determined, since this calculation is performed after all the natural frequencies have been computed. This number is specified in the seventh field of the EIGR card (ND). Six normal modes are requested to be determined in this example.

15.5 Results

Exhibits 15.6 shows the natural frequencies. The first five values coincide with those obtained with the inverse power method (Exhibit 14.6). Computation of the GENERALIZED MASS and GENERALIZED STIFFNESS requires the knowledge of both the natural frequency and the normal mode. Therefore, these quantities are computed only for the first six modes, consistent with the choice of six eigenvectors made on the EIGR card.

The normal modes (Exhibit 15.7) again coincide with those of Problem 14 (Exhibit 14.7). Some of them, however, have been scaled differently -- the second mode, for instance. This is because the unit mass criterion for scaling the eigenvectors, namely $\{\phi\}^T[M]\{\phi\}=1$, does not uniquely determine the sign of the components of $\{\phi\}$.

Finally, considering that the beam model of Figure 14.2 has nine GRIDs and two degrees of freedom per GRID (displacement in the y-direction and rotation about z) a total of $9 \times 2 = 18$ degrees of freedom are obtained. Therefore, one might have expected eighteen natural frequencies to be determined. Why do we have only nine natural frequencies then?

The reason is that the default mass matrix (lumped mass matrix) is used. Recall that this matrix considers only the translational degrees of freedom. There are only nine translational degrees of freedom in this model, which results in a 9×9 mass matrix. For this reason the size of the eigenvalue problem is reduced and accordingly only nine natural frequencies are determined. (Exhibit 15.5 shows the singularity table for the mass matrix which indicates that rotational degrees of freedom have no inertia terms associated with them).

```
N A S T R A N   E X E C U T I V E   C O N T R O L   D E C K   E C H O

    ID   A,B
    TIME  5
    SOL   3
    CEND
```

Exhibit 15.1   Executive Control Deck

```
                  C A S E    C O N T R O L   D E C K   E C H O
CARD
COUNT
  1      TITLE=     NORMAL  MODES AND  NAT. FREQ.
  2      SUBTITLE=  USING   MODIFIED GIVENS METHOD
  3      DISP=ALL
  4      ECHO=BOTH
  5      METHOD= 99
  6      BEGIN BULK
```

Exhibit 15.2   Case Control Deck

```
                    I N P U T   B U L K   D A T A   D E C K   E C H O

 .  1  ..  2  ..  3  ..  4  ..  5  ..  6  ..  7  ..  8  ..  9
$$
GRID         101            00.00   0.00   0.00         1345
GRID         102            5.      0.00   0.00         1345
GRID         103            10.00   0.00   0.00         1345
GRID         104            15.     0.00   0.00         1345
GRID         105            20.00   0.00   0.00         1345
GRID         106            25.00   0.00   0.00         1345
GRID         107            30.00   0.00   0.00         1345
GRID         108            35.00   0.00   0.00         1345
GRID         109            40.     0.00   0.00         1345
$$
CBAR         1001 5555      101     102    1.000  0.000  1.000
CBAR         1002 5555      102     103    1.000  0.000  1.000
CBAR         1003 5555      103     104    1.000  0.000  1.000
CBAR         1004 5555      104     105    1.000  0.000  1.000
CBAR         1005 5555      105     106    1.000  0.000  1.000
CBAR         1006 5555      106     107    1.000  0.000  1.000
CBAR         1007 5555      107     108    1.000  0.000  1.000
CBAR         1008 5555      108     109    1.000  0.000  1.000
$$
PBAR    5555    8888    4.00            1.333
MAT1    8888    30.0+6           .29    7.0-4
$$
$$  CONCENTRATED MASS   AT POINT  102
$$
CMASS2  123456  .02      102      2
$$
$$ SPRINGS AT NODE  101    AND   109
$$
CELAS2  2233    10.+4   101      2
CELAS2  2244    10.+4   109      2
$$ NOW    WE   SELECT   THE MODIFIED GIVENS  METHOD
EIGR    99      MGIV                           6
ENDDATA
```

Exhibit 15.3   Input Bulk Data Deck

CARD COUNT	. 1	.. 2	.. 3	.. 4	.. 5	.. 6	.. 7	.. 8
	S O R T E D		B U L K		D A T A		E C H O	
1-	CBAR	1001	5555	101	102	1.000	0.000	1.000
2-	CBAR	1002	5555	102	103	1.000	0.000	1.000
3-	CBAR	1003	5555	103	104	1.000	0.000	1.000
4-	CBAR	1004	5555	104	105	1.000	0.000	1.000
5-	CBAR	1005	5555	105	106	1.000	0.000	1.000
6-	CBAR	1006	5555	106	107	1.000	0.000	1.000
7-	CBAR	1007	5555	107	108	1.000	0.000	1.000
8-	CBAR	1008	5555	108	109	1.000	0.000	1.000
9-	CELAS2	2233	10.+4	101	2			
10-	CELAS2	2244	10.+4	109	2			
11-	CMASS2	123456	.02	102	2			
12-	EIGR	99	MGIV			6		
13-	GRID	101		00.00	0.00	0.00		1345
14-	GRID	102		5.	0.00	0.00		1345
15-	GRID	103		10.00	0.00	0.00		1345
16-	GRID	104		15.	0.00	0.00		1345
17-	GRID	105		20.00	0.00	0.00		1345
18-	GRID	106		25.00	0.00	0.00		1345
19-	GRID	107		30.00	0.00	0.00		1345
20-	GRID	108		35.00	0.00	0.00		1345
21-	GRID	109		40.	0.00	0.00		1345
22-	MAT1	8888	30.0+6		.29	7.0-4		
23-	PBAR	5555	8888	4.00		1.333		
	ENDDATA							

Exhibit 15.4   Bulk Data Deck sorted by alphabetical order

SEQUENCE PROCESSOR OUTPUT

COLUMN 1

VAXW

POINT	VALUE	POINT	VALUE	POINT	VALUE	POINT	VALUE	POINT	VALUE
101 R3	1.00000E+00	102 R3	1.00000E+00	103 R3	1.00000E+00	104 R3	1.00000E+00	105 R3	1.00000E+00
106 R3	1.00000E+00	107 R3	1.00000E+00	108 R3	1.00000E+00	109 R3	1.00000E+00		

Exhibit 15.5   Sequence processor output

R E A L   E I G E N V A L U E S

MODE NO.	EXTRACTION ORDER	EIGENVALUE	RADIANS	CYCLES	GENERALIZED MASS	GENERALIZED STIFFNESS
1	1	4.011898E+05	6.333954E+02	1.008080E+02	1.000000E+00	4.011898E+05
2	2	2.870254E+06	1.694182E+03	2.696375E+02	1.000000E+00	2.870254E+06
3	3	8.406914E+06	2.899468E+03	4.614646E+02	1.000000E+00	8.406914E+06
4	4	2.469761E+07	4.969669E+03	7.909474E+02	1.000000E+00	2.469761E+07
5	5	6.807676E+07	8.250864E+03	1.313166E+03	1.000000E+00	6.807676E+07
6	6	1.525866E+08	1.235259E+04	1.965976E+03	1.000000E+00	1.525866E+08
7	7	3.208477E+08	1.791222E+04	2.850819E+03	0.0	0.0
8	8	6.047239E+08	2.459113E+04	3.913801E+03	0.0	0.0
9	9	9.357556E+08	3.059012E+04	4.868569E+03	0.0	0.0

Exhibit 15.6   Eigenvalues

EIGENVALUE = 4.011898E+05
CYCLES = 1.008080E+02

REAL EIGENVECTOR NO. 1

POINT ID.	TYPE	T1	T2	T3	R1	R2	R3
101	G	0.0	7.474706E-01	0.0	0.0	0.0	2.532281E-01
102	G	0.0	1.975764E+00	0.0	0.0	0.0	2.305199E-01
103	G	0.0	2.991016E+00	0.0	0.0	0.0	1.708196E-01
104	G	0.0	3.645098E+00	0.0	0.0	0.0	8.780219E-02
105	G	0.0	3.849594E+00	0.0	0.0	0.0	-6.881504E-03
106	G	0.0	3.578769E+00	0.0	0.0	0.0	-1.000735E-01
107	G	0.0	2.873081E+00	0.0	0.0	0.0	-1.787323E-01
108	G	0.0	1.834547E+00	0.0	0.0	0.0	-2.315307E-01
109	G	0.0	6.146500E-01	0.0	0.0	0.0	-2.502037E-01

EIGENVALUE = 2.870254E+06
CYCLES = 2.696375E+02

REAL EIGENVECTOR NO. 2

POINT ID.	TYPE	T1	T2	T3	R1	R2	R3
101	G	0.0	3.508428E+00	0.0	0.0	0.0	-3.508443E-03
102	G	0.0	3.344832E+00	0.0	0.0	0.0	-9.114043E-02
103	G	0.0	2.474969E+00	0.0	0.0	0.0	-2.520054E-01
104	G	0.0	9.245867E-01	0.0	0.0	0.0	-3.529855E-01
105	G	0.0	-8.839109E-01	0.0	0.0	0.0	-3.513805E-01
106	G	0.0	-2.417392E+00	0.0	0.0	0.0	-2.466797E-01
107	G	0.0	-3.247991E+00	0.0	0.0	0.0	-8.034893E-02
108	G	0.0	-3.236765E+00	0.0	0.0	0.0	7.645167E-02
109	G	0.0	-2.635111E+00	0.0	0.0	0.0	1.422703E-01

Exhibit 15.7 Eigenvectors

EIGENVALUE = 8.406914E+06
CYCLES   = 4.614646E+02

REAL EIGENVECTOR NO. 3

POINT ID.	TYPE	T1	T2	T3	R1	R2	R3
101	G	0.0	-4.349828E+00	0.0	0.0	0.0	4.432199E-01
102	G	0.0	-2.040475E+00	0.0	0.0	0.0	4.991722E-01
103	G	0.0	5.245544E-01	0.0	0.0	0.0	4.847210E-01
104	G	0.0	2.417713E+00	0.0	0.0	0.0	2.368567E-01
105	G	0.0	2.684507E+00	0.0	0.0	0.0	-1.361764E-01
106	G	0.0	1.160181E+00	0.0	0.0	0.0	-4.466704E-01
107	G	0.0	-1.442226E+00	0.0	0.0	0.0	-5.531815E-01
108	G	0.0	-4.048956E+00	0.0	0.0	0.0	-4.660858E-01
109	G	0.0	-6.117102E+00	0.0	0.0	0.0	-3.874010E-01

EIGENVALUE = 2.469761E+07
CYCLES   = 7.909474E+02

REAL EIGENVECTOR NO. 4

POINT ID.	TYPE	T1	T2	T3	R1	R2	R3
101	G	0.0	-5.540372E+00	0.0	0.0	0.0	1.039024E+00
102	G	0.0	-5.556172E-01	0.0	0.0	0.0	9.128048E-01
103	G	0.0	2.923883E+00	0.0	0.0	0.0	3.883100E-01
104	G	0.0	2.947312E+00	0.0	0.0	0.0	-3.642871E-01
105	G	0.0	-4.156893E-02	0.0	0.0	0.0	-7.104329E-01
106	G	0.0	-2.956182E+00	0.0	0.0	0.0	-3.360775E-01
107	G	0.0	-2.741498E+00	0.0	0.0	0.0	4.347852E-01
108	G	0.0	1.026182E+00	0.0	0.0	0.0	9.863549E-01
109	G	0.0	6.447579E+00	0.0	0.0	0.0	1.133242E+00

Exhibit 15.7 (continued)   Eigenvectors

EIGENVALUE = 6.807676E+07
CYCLES = 1.313166E+03

REAL EIGENVECTOR NO. 5

POINT ID.	TYPE	T1	T2	T3	R1	R2	R3
101	G	0.0	6.487302E+00	0.0	0.0	0.0	-1.788923E+00
102	G	0.0	-1.184751E+00	0.0	0.0	0.0	-1.025385E+00
103	G	0.0	-2.650028E+00	0.0	0.0	0.0	4.080660E-01
104	G	0.0	1.268041E+00	0.0	0.0	0.0	8.647964E-01
105	G	0.0	3.683887E+00	0.0	0.0	0.0	-6.690235E-02
106	G	0.0	7.436863E-01	0.0	0.0	0.0	-9.117998E-01
107	G	0.0	-3.091076E+00	0.0	0.0	0.0	-3.508763E-01
108	G	0.0	-1.587561E+00	0.0	0.0	0.0	9.165565E-01
109	G	0.0	4.928981E+00	0.0	0.0	0.0	1.496684E+00

EIGENVALUE = 1.525866E+08
CYCLES = 1.965976E+03

REAL EIGENVECTOR NO. 6

POINT ID.	TYPE	T1	T2	T3	R1	R2	R3
101	G	0.0	-5.179598E+00	0.0	0.0	0.0	1.961943E+00
102	G	0.0	2.017795E+00	0.0	0.0	0.0	3.945505E-01
103	G	0.0	-1.005178E+00	0.0	0.0	0.0	-1.035493E+00
104	G	0.0	-3.773315E+00	0.0	0.0	0.0	2.727547E-01
105	G	0.0	9.682558E-01	0.0	0.0	0.0	1.128534E+00
106	G	0.0	3.636195E+00	0.0	0.0	0.0	-3.411857E-01
107	G	0.0	-1.195192E+00	0.0	0.0	0.0	-1.061860E+00
108	G	0.0	-3.017433E+00	0.0	0.0	0.0	5.964496E-01
109	G	0.0	4.046093E+00	0.0	0.0	0.0	1.820833E+00

Exhibit 15.7 (continued) Eigenvectors

16
Problem 16

16.1 Statement of the problem

Calculate the normal modes and natural frequencies of the beam structure analyzed in Problems 14 and 15 (Figures 14.1 and 14.2) using the coupled mass matrix option.[1]

16.2 Cards introduced

Case Control Deck   None

Bulk Data Deck   PARAM,COUPMASS,+1

16.3 MSC/NASTRAN formulation

The modified Givens method is employed to extract the eigenvalues (as in the previous example), but the coupled mass matrix option is chosen instead of the default lumped mass matrix option. Notice that all the eigenvalue extraction algorithms can be used with either type of mass matrix.

The coupled mass matrix incorporates the inertial effects of the rotational degrees of freedom, and includes some off-diagonal terms. It is generally more accurate than the lumped mass matrix but more expensive to use from a numerical standpoint.

---

[1] The MSC/NASTRAN coupled mass matrix is a compromise between the lumped mass matrix and the classical consistent mass matrix.

16.4 Input Data Deck

Exhibit 16.1 shows the Executive Control Deck and Exhibit 16.2 shows the Case Control Deck.

The Bulk Data Deck (Exhibits 16.3 and 16.4) is similar to that of Problem 15 except for two minor modifications. First, a special statement is included to request that the coupled mass matrix be used. This statement is PARAM,COUPMASS,+1 and can go anywhere in the Bulk Data Deck. Second, in this example the MAX option is selected for scaling the normal modes instead of the MASS (default) option that was used in Problem 14. This option is selected on the EIGR card (first field of the continuation card). The MAX option means that the normal modes are scaled in such a way that the largest component of displacement (in absolute value) has unit magnitude. The scaling option chosen for the normal modes is independent of both the mass matrix used and the eigenvalue extraction technique employed.

16.5 Results

The natural frequencies are presented in Exhibit 16.5. Note that while eighteen natural frequencies are calculated in this problem, only nine were computed with the lumped mass matrix approach (Exhibit 15.6). This is because the coupled mass matrix takes into account the inertial effects of the rotational degrees of freedom, which results in an 18×18 mass matrix (in contrast to 9×9 in the previous example). Therefore, the size of the eigenvalue problem is eighteen.

Notice also that 1's do not appear under GENERALIZED MASS (Exhibit 16.5) since a different criterion was selected to scale the normal modes (MAX instead of MASS). The ratio between GENERALIZED STIFFNESS and GENERALIZED MASS is, however, still equal to the corresponding EIGENVALUE.

The results for the frequencies are similar to those obtained in Problem 15 for the first few frequencies. Some differences begin to appear in the high frequency modes since rotational degrees of freedom play a more important role in such modes, and therefore the choice of mass matrix becomes a more important factor.

Finally, Exhibit 16.6 shows the normal modes with the maximum displacement component normalized to one. These normal modes roughly correspond to the same deformation patterns given by the eigenvectors of the previous problem (Exhibit 15.7); some discrepancies are more prominent in the higher modes.

```
N A S T R A N   E X E C U T I V E   C O N T R O L   D E C K   E C H O

    ID   A,B
    TIME  5
    SOL   3
    CEND
```

Exhibit 16.1   Executive Control Deck

```
                C A S E    C O N T R O L   D E C K   E C H O
CARD
COUNT
  1     TITLE=     NORMAL  MODES AND  NAT. FREQ.
  2     SUBTITLE=  USING  COUPLED MASS  MATRIX
  3     DISP=ALL
  4     ECHO=BOTH
  5     METHOD= 99
  6     BEGIN BULK
```

Exhibit 16.2   Case Control Deck

```
                    I N P U T   B U L K   D A T A   D E C K   E C H O
      .  ..  1  ..  2  ..  3  ..  4  ..  5  ..  6  ..  7  ..  8  ..  9  ..  10  .
     $$
     GRID       101           00.00   0.00   0.00   0.00   1345
     GRID       102           5.      0.00   0.00   0.00   1345
     GRID       103           10.00   0.00   0.00   0.00   1345
     GRID       104           15.     0.00   0.00   0.00   1345
     GRID       105           20.00   0.00   0.00   0.00   1345
     GRID       106           25.00   0.00   0.00   0.00   1345
     GRID       107           30.00   0.00   0.00   0.00   1345
     GRID       108           35.00   0.00   0.00   0.00   1345
     GRID       109           40.      0.00   0.00   0.00   1345
     $$
     CBAR      1001   5555    101    102    1.000  0.000  1.000
     CBAR      1002   5555    102    103    1.000  0.000  1.000
     CBAR      1003   5555    103    104    1.000  0.000  1.000
     CBAR      1004   5555    104    105    1.000  0.000  1.000
     CBAR      1005   5555    105    106    1.000  0.000  1.000
     CBAR      1006   5555    106    107    1.000  0.000  1.000
     CBAR      1007   5555    107    108    1.000  0.000  1.000
     CBAR      1008   5555    108    109    1.000  0.000  1.000
     $$
     PBAR      5555   8888    4.00          1.333
     MAT1      8888   30.0+6          .29   7.0-4
     $$
```

Exhibit 16.3 Input Bulk Data Deck

```
$$
$$   CONCENTRATED MASS     AT POINT   102
$$
CMASS2  123456  .02      102     2
$$
$$  SPRINGS AT NODE  101     AND   109
$$
CELAS2  2233    10.+4    101     2
CELAS2  2244    10.+4    109     2
$$
$$   NOW WE NORMALIZE THE EIGENVECTORS USING THE     MAX  OPTION
$$
EIGR    99      MGIV                    6                               +CCCC
+CCCC   MAX
$$
$$  THIS CARD IS TO USE THE COUPLED MASS MATRIX
$$
PARAM,COUPMASS,+1
$$
ENDDATA
```

Exhibit 16.3 (continued) Input Bulk Data Deck

```
                              S O R T E D   B U L K   D A T A   E C H O

CARD
COUNT    .    1  ..   2   ..   3   ..  4   ..  5   ..  6   ..  7   ..  8   ..  9  ..  10  .
 1-    CBAR     1001    5555    101    102   1.000   0.000   1.000
 2-    CBAR     1002    5555    102    103   1.000   0.000   1.000
 3-    CBAR     1003    5555    103    104   1.000   0.000   1.000
 4-    CBAR     1004    5555    104    105   1.000   0.000   1.000
 5-    CBAR     1005    5555    105    106   1.000   0.000   1.000
 6-    CBAR     1006    5555    106    107   1.000   0.000   1.000
 7-    CBAR     1007    5555    107    108   1.000   0.000   1.000
 8-    CBAR     1008    5555    108    109   1.000   0.000   1.000
 9-    CELAS2   2233    10.+4   101    2
10-    CELAS2   2244    10.+4   109    2
11-    CMASS2   123456  .02     102    2
12-    EIGR     99      MGIV
13-    +CCCC    MAX                          6                              +CCCC
```

Exhibit 16.4   Bulk Data Deck sorted by alphabetical order

```
                           S O R T E D   B U L K   D A T A   E C H O

CARD
COUNT    .   1   ..   2   ..   3   ..   4   ..   5   ..   6   ..   7   ..   8   ..   9   ..   10  .
14-    GRID      101            00.00    0.00    0.00          1345
15-    GRID      102            5.       0.00    0.00          1345
16-    GRID      103            10.00    0.00    0.00          1345
17-    GRID      104            15.      0.00    0.00          1345
18-    GRID      105            20.00    0.00    0.00          1345
19-    GRID      106            25.00    0.00    0.00          1345
20-    GRID      107            30.00    0.00    0.00          1345
21-    GRID      108            35.00    0.00    0.00          1345
22-    GRID      109            40.      0.00    0.00          1345
23-    MAT1      8888   30.0+6           .29     7.0-4
24-    PARAM   COUPMASS+1
25-    PBAR      5555   8888    4.00            1.333
       ENDDATA
```

Exhibit 16.4 (continued)   Bulk Data Deck sorted by alphabetical order

MODE NO.	EXTRACTION ORDER	EIGENVALUE	REAL EIGENVALUES		GENERALIZED MASS	GENERALIZED STIFFNESS
			RADIANS	CYCLES		
1	1	3.996023E+05	6.321410E+02	1.006084E+02	6.780570E-02	2.709531E+04
2	2	2.857538E+06	1.690425E+03	2.690395E+02	8.417942E-02	2.405458E+05
3	3	8.824535E+06	2.970612E+03	4.727875E+02	2.721264E-02	2.401389E+05
4	4	2.888959E+07	5.374904E+03	8.554425E+02	2.074107E-02	5.992011E+05
5	5	8.816048E+07	9.389381E+03	1.494366E+03	1.783927E-02	1.572718E+06
6	6	2.082706E+08	1.443158E+04	2.296858E+03	1.519615E-02	3.164911E+06
7	7	4.392664E+08	2.095868E+04	3.335678E+03	0.0	0.0
8	8	8.866637E+08	2.977690E+04	4.739141E+03	0.0	0.0
9	9	1.651402E+09	4.063744E+04	6.467649E+03	0.0	0.0
10	10	3.376841E+09	5.811059E+04	9.248587E+03	0.0	0.0
11	11	5.493735E+09	7.411973E+04	1.179652E+04	0.0	0.0
12	12	8.711903E+09	9.333758E+04	1.485514E+04	0.0	0.0
13	13	1.314075E+10	1.146331E+05	1.824443E+04	0.0	0.0
14	14	1.985121E+10	1.408943E+05	2.242403E+04	0.0	0.0
15	15	3.068282E+10	1.751651E+05	2.787839E+04	0.0	0.0
16	16	4.608331E+10	2.146703E+05	3.416583E+04	0.0	0.0
17	17	7.710505E+10	2.776780E+05	4.419382E+04	0.0	0.0
18	18	8.204997E+10	2.864437E+05	4.558892E+04	0.0	0.0

Exhibit 16.5 Eigenvalues

EIGENVALUE = 3.996023E+05
CYCLES = 1.006084E+02

REAL EIGENVECTOR NO. 1

POINT ID.	TYPE	T1	T2	T3	R1	R2	R3
101	G	0.0	1.950445E-01	0.0	0.0	0.0	6.576263E-02
102	G	0.0	5.138863E-01	0.0	0.0	0.0	5.982727E-02
103	G	0.0	7.773020E-01	0.0	0.0	0.0	4.431741E-02
104	G	0.0	9.469633E-01	0.0	0.0	0.0	2.277488E-02
105	G	0.0	1.000000E+00	0.0	0.0	0.0	-1.784965E-03
106	G	0.0	9.297612E-01	0.0	0.0	0.0	-2.595730E-02
107	G	0.0	7.467225E-01	0.0	0.0	0.0	-4.636739E-02
108	G	0.0	4.772888E-01	0.0	0.0	0.0	-6.008343E-02
109	G	0.0	1.606704E-01	0.0	0.0	0.0	-6.496489E-02

EIGENVALUE = 2.857538E+06
CYCLES = 2.690395E+02

REAL EIGENVECTOR NO. 2

POINT ID.	TYPE	T1	T2	T3	R1	R2	R3
101	G	0.0	1.000000E+00	0.0	0.0	0.0	2.479491E-03
102	G	0.0	9.655163E-01	0.0	0.0	0.0	-2.460798E-02
103	G	0.0	7.196475E-01	0.0	0.0	0.0	-7.236572E-02
104	G	0.0	2.717959E-01	0.0	0.0	0.0	-1.024594E-01
105	G	0.0	-2.541341E-01	0.0	0.0	0.0	-1.024866E-01
106	G	0.0	-7.018896E-01	0.0	0.0	0.0	-7.222646E-02
107	G	0.0	-9.453565E-01	0.0	0.0	0.0	-2.363823E-02
108	G	0.0	-9.419035E-01	0.0	0.0	0.0	2.267210E-02
109	G	0.0	-7.619281E-01	0.0	0.0	0.0	4.309614E-02

Exhibit 16.6   Eigenvectors

EIGENVALUE = 8.824535E+06
CYCLES = 4.727875E+02

REAL EIGENVECTOR NO. 3

POINT ID.	TYPE	T1	T2	T3	R1	R2	R3
101	G	0.0	7.249067E-01	0.0	0.0	0.0	-6.916602E-02
102	G	0.0	3.518445E-01	0.0	0.0	0.0	-8.363026E-02
103	G	0.0	-8.717498E-02	0.0	0.0	0.0	-8.462813E-02
104	G	0.0	-4.207869E-01	0.0	0.0	0.0	-4.249688E-02
105	G	0.0	-4.721081E-01	0.0	0.0	0.0	2.312422E-02
106	G	0.0	-2.084147E-01	0.0	0.0	0.0	7.771368E-02
107	G	0.0	2.411652E-01	0.0	0.0	0.0	9.496058E-02
108	G	0.0	6.790656E-01	0.0	0.0	0.0	7.613006E-02
109	G	0.0	1.000000E+00	0.0	0.0	0.0	5.679778E-02

EIGENVALUE = 2.888959E+07
CYCLES = 8.554425E+02

REAL EIGENVECTOR NO. 4

POINT ID.	TYPE	T1	T2	T3	R1	R2	R3
101	G	0.0	-8.243254E-01	0.0	0.0	0.0	1.413130E-01
102	G	0.0	-1.107273E-01	0.0	0.0	0.0	1.398600E-01
103	G	0.0	4.445496E-01	0.0	0.0	0.0	6.628387E-02
104	G	0.0	4.646565E-01	0.0	0.0	0.0	-5.708929E-02
105	G	0.0	-1.320648E-02	0.0	0.0	0.0	-1.148397E-01
106	G	0.0	-4.712520E-01	0.0	0.0	0.0	-4.965993E-02
107	G	0.0	-4.041602E-01	0.0	0.0	0.0	7.695056E-02
108	G	0.0	2.102096E-01	0.0	0.0	0.0	1.538574E-01
109	G	0.0	1.000000E+00	0.0	0.0	0.0	1.564095E-01

Exhibit 16.6 (continued) Eigenvectors

EIGENVALUE =   8.816048E+07
CYCLES =   1.494366E+03

REAL EIGENVECTOR NO.   5

POINT ID.	TYPE	T1	T2	T3	R1	R2	R3
101	G	0.0	1.000000E+00	0.0	0.0	0.0	-2.342381E-01
102	G	0.0	-1.003901E-01	0.0	0.0	0.0	-1.746450E-01
103	G	0.0	-4.504861E-01	0.0	0.0	0.0	4.144514E-02
104	G	0.0	1.142215E-01	0.0	0.0	0.0	1.421645E-01
105	G	0.0	5.320505E-01	0.0	0.0	0.0	-4.138747E-03
106	G	0.0	8.321756E-02	0.0	0.0	0.0	-1.446992E-01
107	G	0.0	-4.763505E-01	0.0	0.0	0.0	-3.887071E-02
108	G	0.0	-1.440173E-01	0.0	0.0	0.0	1.594955E-01
109	G	0.0	8.359044E-01	0.0	0.0	0.0	2.074492E-01

EIGENVALUE =   2.082706E+08
CYCLES =   2.296858E+03

REAL EIGENVECTOR NO.   6

POINT ID.	TYPE	T1	T2	T3	R1	R2	R3
101	G	0.0	1.000000E+00	0.0	0.0	0.0	-2.903237E-01
102	G	0.0	-2.342164E-01	0.0	0.0	0.0	-1.272087E-01
103	G	0.0	-1.909247E-02	0.0	0.0	0.0	1.524016E-01
104	G	0.0	4.914604E-01	0.0	0.0	0.0	-8.862217E-03
105	G	0.0	-9.512581E-02	0.0	0.0	0.0	-1.650744E-01
106	G	0.0	-4.551510E-01	0.0	0.0	0.0	5.854297E-02
107	G	0.0	2.306992E-01	0.0	0.0	0.0	1.414401E-01
108	G	0.0	3.109792E-01	0.0	0.0	0.0	-1.284593E-01
109	G	0.0	-7.269518E-01	0.0	0.0	0.0	-2.372410E-01

Exhibit 16.6 (continued) Eigenvectors

17
Problem 17

17.1 Statement of the problem

Compute the normal modes and natural frequencies of the beam structure of Problem 14 (Figures 14.1 and 14.2) assuming that the springs at both ends are removed. Thus, it is now a free (unconstrained) structure.

17.2 Cards introduced

Case Control Deck   None

Bulk Data Deck   None

17.3 MSC/NASTRAN formulation

The model used is the same as Problems 14, 15, and 16 (Figure 14.2) except that the CELAS2 cards have been deleted. The modified Givens method and the lumped mass matrix option (default) are employed. Any other eigenvalue extraction method would have been equally acceptable.

17.4 Input Data Deck

The Executive Control Deck and the Case Control Deck, shown in Exhibits 17.1 and 17.2 respectively, should be familiar at this point.

The Bulk Data Deck (Exhibit 17.3) is similar to that of Problem 15 except for two minor changes. First, the CELAS2 cards used to specify the location and properties of the springs have been deleted. Second, the EIGR card specifies a value of ND=4, requesting that only the first four normal modes be calculated.

PARAM,COUPMASS,+1 is not included which means that the lumped mass matrix will be used in the eigenvalue calculation.

17.5 Results

Exhibit 17.5 shows the singularity table which indicates that rotational degrees of freedom have no stiffness associated with them. This is consistent with the fact that the lumped mass matrix is used.

The natural frequencies are listed in Exhibit 17.6. The first two frequencies are of the order of 10^{-6}, that is, computational zeros. This is to be expected since this is a free structure in a two-dimensional space, and it can move in the y-direction as well as rotate about z. If the x-displacement had not been constrained on each GRID card, the beam would have exhibited three zero natural frequencies. (Six zero-frequencies would have been obtained in a general three-dimensional problem regardless of the type of mass matrix used). Each zero-frequency is associated with a normal mode that represents a rigid body (stress-free) mode.

The normal modes are presented in Exhibit 17.7. The first two normal modes correspond to rigid body modes. In general, the rigid body modes will not correspond to a purely translational mode or a pure-rotational mode. These modes, however, when appropriately combined, can represent a pure-rotation configuration and a pure translation configuration.

Finally, notice that the first non-zero frequency (244.88 Hz) is different than the first frequency of the same beam when it was supported on two springs (100.80 Hz, from Problem 14, Exhibit 14.6). These two values are not related in any way!

```
N A S T R A N   E X E C U T I V E   C O N T R O L   D E C K   E C H O

      ID   A,B
      TIME  5
      SOL   3
      CEND
```

Exhibit 17.1 Executive Control Deck

```
                  C A S E    C O N T R O L    D E C K    E C H O
CARD
COUNT
  1       TITLE=     NORMAL  MODES AND  NAT. FREQ.
  2       SUBTITLE=  OF  A FREE STRUCTURE
  3       DISP=ALL
  4       ECHO=BOTH
  5       METHOD= 99
  6       BEGIN BULK
```

Exhibit 17.2   Case Control Deck

```
                  I N P U T    B U L K    D A T A    D E C K    E C H O
  .   1  ..   2  ..   3  ..   4  ..   5  ..   6  ..   7  ..   8  ..   9
GRID        101          00.00   0.00   0.00          1345
GRID        102          5.      0.00   0.00          1345
GRID        103          10.00   0.00   0.00          1345
GRID        104          15.     0.00   0.00          1345
GRID        105          20.00   0.00   0.00          1345
GRID        106          25.00   0.00   0.00          1345
GRID        107          30.00   0.00   0.00          1345
GRID        108          35.00   0.00   0.00          1345
GRID        109          40.     0.00   0.00          1345
$$
CBAR        1001 5555       101    102   1.000  0.000  1.000
CBAR        1002 5555       102    103   1.000  0.000  1.000
CBAR        1003 5555       103    104   1.000  0.000  1.000
CBAR        1004 5555       104    105   1.000  0.000  1.000
CBAR        1005 5555       105    106   1.000  0.000  1.000
CBAR        1006 5555       106    107   1.000  0.000  1.000
CBAR        1007 5555       107    108   1.000  0.000  1.000
CBAR        1008 5555       108    109   1.000  0.000  1.000
$$
PBAR    5555    8888    4.00            1.333
MAT1    8888    30.0+6           .29    7.0-4
$$ CONCENTRATED MASS   AT POINT  102
CMASS2  123456  .02     102     2
$$
EIGR    99      MGIV                           4
ENDDATA
```

Exhibit 17.3   Input Bulk Data Deck

```
                        S O R T E D   B U L K   D A T A   E C H O
CARD
COUNT    .  1  ..  2  ..  3  ..  4  ..  5  ..  6  ..  7  ..  8
   1-    CBAR    1001    5555    101     102     1.000   0.000   1.000
   2-    CBAR    1002    5555    102     103     1.000   0.000   1.000
   3-    CBAR    1003    5555    103     104     1.000   0.000   1.000
   4-    CBAR    1004    5555    104     105     1.000   0.000   1.000
   5-    CBAR    1005    5555    105     106     1.000   0.000   1.000
   6-    CBAR    1006    5555    106     107     1.000   0.000   1.000
   7-    CBAR    1007    5555    107     108     1.000   0.000   1.000
   8-    CBAR    1008    5555    108     109     1.000   0.000   1.000
   9-    CMASS2  123456  .02     102     2
  10-    EIGR    99      MGIV                                    4
  11-    GRID    101             00.00   0.00    0.00                    1345
  12-    GRID    102             5.      0.00    0.00                    1345
  13-    GRID    103             10.00   0.00    0.00                    1345
  14-    GRID    104             15.     0.00    0.00                    1345
  15-    GRID    105             20.00   0.00    0.00                    1345
  16-    GRID    106             25.00   0.00    0.00                    1345
  17-    GRID    107             30.00   0.00    0.00                    1345
  18-    GRID    108             35.00   0.00    0.00                    1345
  19-    GRID    109             40.     0.00    0.00                    1345
  20-    MAT1    8888    30.0+6          .29     7.0-4
  21-    PBAR    5555    8888    4.00            1.333
         ENDDATA
```

Exhibit 17.4   Bulk Data Deck sorted by alphabetical order

SEQUENCE PROCESSOR OUTPUT

VAXW

COLUMN   1

POINT	VALUE	POINT	VALUE	POINT	VALUE	POINT	VALUE	POINT	VALUE
101 R3	1.00000E+00	102 R3	1.00000E+00	103 R3	1.00000E+00	104 R3	1.00000E+00	105 R3	1.00000E+00
106 R3	1.00000E+00	107 R3	1.00000E+00	108 R3	1.00000E+00	109 R3	1.00000E+00		

Exhibit 17.5   Sequence processor output

REAL EIGENVALUES

MODE NO.	EXTRACTION ORDER	EIGENVALUE	RADIANS	CYCLES	GENERALIZED MASS	GENERALIZED STIFFNESS
1	1	2.328306E-10	1.525879E-05	2.428512E-06	1.000000E+00	2.328306E-10
2	9	-4.656613E-10	2.157919E-05	3.434434E-06	1.000000E+00	-4.656613E-10
3	2	2.367446E+06	1.538651E+03	2.448839E+02	1.000000E+00	2.367446E+06
4	3	1.818788E+07	4.264726E+03	6.787521E+02	1.000000E+00	1.818788E+07
5	4	6.198549E+07	7.873086E+03	1.253041E+03	0.0	0.0
6	5	1.485616E+08	1.218859E+04	1.939874E+03	0.0	0.0
7	6	3.190312E+08	1.786145E+04	2.842737E+03	0.0	0.0
8	7	6.040547E+08	2.457753E+04	3.911635E+03	0.0	0.0
9	8	9.356067E+08	3.058769E+04	4.868182E+03	0.0	0.0

Exhibit 17.6   Eigenvalues

EIGENVALUE = 2.328306E-10
CYCLES = 2.428512E-06

REAL EIGENVECTOR NO. 1

POINT ID.	TYPE	T1	T2	T3	R1	R2	R3
101	G	0.0	-4.789010E+00	0.0	0.0	0.0	1.584678E-01
102	G	0.0	-3.996670E+00	0.0	0.0	0.0	1.584678E-01
103	G	0.0	-3.204331E+00	0.0	0.0	0.0	1.584678E-01
104	G	0.0	-2.411992E+00	0.0	0.0	0.0	1.584678E-01
105	G	0.0	-1.619653E+00	0.0	0.0	0.0	1.584678E-01
106	G	0.0	-8.273137E-01	0.0	0.0	0.0	1.584678E-01
107	G	0.0	-3.497455E-02	0.0	0.0	0.0	1.584678E-01
108	G	0.0	7.573646E-01	0.0	0.0	0.0	1.584678E-01
109	G	0.0	1.549704E+00	0.0	0.0	0.0	1.584678E-01

EIGENVALUE = -4.656613E-10
CYCLES = 3.434434E-06

REAL EIGENVECTOR NO. 2

POINT ID.	TYPE	T1	T2	T3	R1	R2	R3
101	G	0.0	9.965791E-01	0.0	0.0	0.0	-1.640793E-01
102	G	0.0	1.761828E-01	0.0	0.0	0.0	-1.640793E-01
103	G	0.0	-6.442136E-01	0.0	0.0	0.0	-1.640793E-01
104	G	0.0	-1.464610E+00	0.0	0.0	0.0	-1.640793E-01
105	G	0.0	-2.285006E+00	0.0	0.0	0.0	-1.640793E-01
106	G	0.0	-3.105402E+00	0.0	0.0	0.0	-1.640793E-01
107	G	0.0	-3.925799E+00	0.0	0.0	0.0	-1.640793E-01
108	G	0.0	-4.746195E+00	0.0	0.0	0.0	-1.640793E-01
109	G	0.0	-5.566591E+00	0.0	0.0	0.0	-1.640793E-01

Exhibit 17.7 Eigenvectors

EIGENVALUE = 2.367446E+06
CYCLES   = 2.448839E+02

REAL EIGENVECTOR NO. 3

POINT ID.	TYPE	T1	T2	T3	R1	R2	R3
101	G	0.0	-4.573084E+00	0.0	0.0	0.0	6.148862E-01
102	G	0.0	-1.538135E+00	0.0	0.0	0.0	5.911972E-01
103	G	0.0	1.195425E+00	0.0	0.0	0.0	4.814302E-01
104	G	0.0	3.088845E+00	0.0	0.0	0.0	2.592700E-01
105	G	0.0	3.674779E+00	0.0	0.0	0.0	-3.089776E-02
106	G	0.0	2.783309E+00	0.0	0.0	0.0	-3.190008E-01
107	G	0.0	5.997239E-01	0.0	0.0	0.0	-5.381322E-01
108	G	0.0	-2.424645E+00	0.0	0.0	0.0	-6.532432E-01
109	G	0.0	-5.790852E+00	0.0	0.0	0.0	-6.832403E-01

EIGENVALUE = 1.818788E+07
CYCLES   = 6.787521E+02

REAL EIGENVECTOR NO. 4

POINT ID.	TYPE	T1	T2	T3	R1	R2	R3
101	G	0.0	-5.336266E+00	0.0	0.0	0.0	1.156465E+00
102	G	0.0	9.212383E-02	0.0	0.0	0.0	9.441035E-01
103	G	0.0	3.426574E+00	0.0	0.0	0.0	3.248252E-01
104	G	0.0	3.146406E+00	0.0	0.0	0.0	-4.108349E-01
105	G	0.0	-3.817945E-03	0.0	0.0	0.0	-7.397210E-01
106	G	0.0	-3.155972E+00	0.0	0.0	0.0	-4.117084E-01
107	G	0.0	-3.445227E+00	0.0	0.0	0.0	3.217092E-01
108	G	0.0	-1.388759E-01	0.0	0.0	0.0	9.351297E-01
109	G	0.0	5.230632E+00	0.0	0.0	0.0	1.143288E+00

Exhibit 17.7 (continued) Eigenvectors

18
Problem 18

18.1 Statement of the problem

Determine the response of the beam structure described in Problem 14 to a time dependent load acting at GRID 105 in the y-direction. Figure 18.1 shows the profile of the excitation, which corresponds to a typical impact load. A dashpot (viscous damper) is connected to each end of the beam (GRIDs 101 and 109). The dashpots have a damping constant $c = 1.5$ lb-s/in and act in the y-direction.

18.2 Cards introduced

Case Control Deck   DLOAD
                                         TSTEP

Bulk Data Deck   CDAMP2
                                         DAREA
                                         TABLED1
                                         TLOAD1
                                         TSTEP

18.3 MSC/NASTRAN formulation

The same model described in Problem 14 and shown in Figure 14.2 is used, except that two dashpots are incorporated using CDAMP2 cards. This type of dynamics problem is called transient analysis. A differential equation of motion of the following form is solved

$$[M]d^2\{x\}/dt^2 + [B]d\{x\}/dt + [K]\{x\} = \{F(t)\} \qquad (18.1)$$

where [M] represents the mass matrix of the structure; [B] represents the damping matrix; [K] represents the stiffness matrix; {x} is the displacement vector; and {F(t)} is the time dependent force (excitation) acting on the structure.

One way of solving this problem is to use a numerical technique to directly integrate the equation of motion in the desired time interval. This can be accomplished in MSC/NASTRAN by invoking in the Executive Control Deck SOL 27, which corresponds to direct transient response. The equation of motion (Eq. 18.1) is integrated using a variation of the well-known Newmark-Beta method. This approach (SOL 27) is demonstrated is this example.

18.4 Input Data Deck

SOL 27 is selected in the Executive Control Deck (Exhibit 18.1). This corresponds to direct transient response.

Examine the Case Control Deck (Exhibit 18.2) where several new cards are introduced. The DLOAD = 100 card selects the Bulk Data Deck card that specifies the dynamic (transient) load acting on the structure. In this case this is a TLOAD1 card. The TSTEP = 200 card is used to select a TSTEP card in the Bulk Data Deck. This card determines certain parameters required for the numerical integration process.

The displacement at GRID 102 is requested to be printed (SET 1000 = 102 followed by DISP = 1000). In this context, the DISP card works almost the same as in linear static analysis (SOL 24). The only difference is that now the displacement is a time dependent variable, and its value is calculated and printed for each time step. The same holds for all the other output request cards (SPCFORCES, OLOAD, STRESS, etc.)

The statements used to create plots to show the response of the structure are quite different than the statements used in previous examples due to the time-dependent nature of this problem.[1] This plot subdeck begins with an OUTPUT(XYPLOT) statement to indicate that these are x-y plots (in MSC/NASTRAN nomenclature), in contrast to SOL 24 and SOL 3 plot subdecks which were preceded by OUTPUT(PLOT). The x-y plots (sometimes called xyplots), show the behavior of a selected recovery variable as a function of time.

The statements between XTAXIS = YES and YBGRID LINES = YES define some general specifications regarding the appearance of the plots.

---

[1] The OUTPUT(PLOT) subdeck can still be used to plot the structure.

They state that the axes should be plotted, subdivisions should be marked along the axes, etc.

Consider now the next group of cards in the plot subdeck. XTITLE = TIME IN SECONDS specifies the title for the horizontal axis of the plot to be created. Analogously, YTITLE = NODE 102 DISPLACEMENT labels the vertical axis. The statement XYPLOT DISP /102(T2) plots the displacement at GRID 102 in the y-direction (T2, T for translation, 2 for the y-direction) as a function of time. The next group of instructions requests that the external load (OLOAD) acting at GRID 105 in the y-direction be plotted as a function of time.

The Bulk Data Deck is shown in Exhibits 18.3 and 18.4. The group of statements at the beginning of the Bulk Data Deck define the dynamic load. The first card is the TLOAD1 card, selected by DLOAD = 100 in the Case Control Deck. The third field of this card points to a DAREA card. The fifth field of the TLOAD1 card contains a 0 indicating that a force as a function of time is being defined, and the number 104 in the sixth field points to a TABLED1 card.

The DAREA card (ID = 101) serves two purposes. First, it specifies the point of application of the load. In this case, it specifies that the load is acting at GRID 105 (this information goes in the third field) and in the y-direction (number 2 in field 4). The second task of this card is to supply a scale factor for the time dependent load. In this example, the scale factor is 58 (field 5).

Finally, TABLED1 (ID = 104) gives a sequence of coordinates pairs that define the time history of the force acting at GRID 105. Since the scale factor is equal to 58 the actual load at any given time will be the value interpolated from this table multiplied by 58.

Two final comments regarding the way the TABLED1 table is set up are pertinent. First, note that at the beginning of the table an extra point is added which means that no load is acting on the structure between 0 and 6×10^{-4} s. That is, the load is applied starting at $t = 6\times10^{-4}$ seconds. What we have actually done is to shift the function shown in Figure 18.1 6×10^{-4} seconds to the right.[2] These few steps with zero-load at the beginning are simply a numerical trick that in general improves the accuracy and stability of the numerical integration scheme. Second, since MSC/NASTRAN always interpolates (or extrapolates) to find the appropriate value from a table, to ensure that a zero force is obtained for values of t larger than 60×10^{-4} s an extra point at the end of the table is added. This point specifies that for a very large t (t = 999999 s) the load is 0. If this extra point is not included, MSC/NASTRAN would simply extrapolate using the last two points of the table, for values of $t > 60\times10^{-4}$ s. This would produce a load approaching minus infinity for large values of t, leading to erroneous results.

---

[2] Practice suggests that the "shift" should be between two to four times the time step selected for the integration process.

The dashpots are specified with two CDAMP2 cards. Each card gives the ID number of the dashpot, the dashpot's viscous damping coefficient, the GRID to which the dashpot is connected, and the degree of freedom affected.

The TSTEP card, selected by TSTEP = 200 in the Case Control Deck, is always required in SOL 27. It specifies the time step for the numerical integration algorithm, in this case $\Delta t = 3 \times 10^{-4}$ s. In addition, the TSTEP card states that 130 integration steps will be carried out (field 3), and that the results will be printed every other time step (2 in field 5).

The selection of an appropriate time step is a crucial factor in the numerical integration process. An excessively large step size might cause the solution to miss some key features of the response (peaks for example), whereas a small step size can make the process very inefficient. A reasonable rule of thumb is to select a step size about 1/10th to 1/7th of the natural period of the highest mode expected to contribute to the solution.

In a previous analysis (Problem 14, Exhibit 14.6) it was determined that the first three natural frequencies of this structure are 100.8 Hz, 269.6 Hz, and 461.4 Hz. Therefore, the corresponding natural periods (inverse of the frequencies) are 0.0099 s, 0.0037 s, and 0.0021 s. The value chosen for the time step (.0003 s) is about 1/7th of the period of the third mode. This is adequate to get a good approximation for a problem of this type since the first few modes are the most likely to dominate the response of the structure. Note also that 0.0003 s is sufficiently small to capture the features of the excitation: there are about eighteen points to describe the profile of the 54×10^{-4} s pulse.

In this example, no specifications regarding the initial conditions are made. Therefore, MSC/NASTRAN assumes that at t = 0, both displacements and velocities are zero for all GRID points. The rest of the Bulk Data Deck is similar to that of Problem 14.

18.5 Results

Exhibit 18.5 shows the displacement at GRID point 102 as a function of time. The first plot (Figure 18.2) corresponds to the displacement at GRID 102 (y-direction) as a function of time. Notice that a rough estimate of the "distance" from peak to peak is about 0.01 s which corresponds to a frequency of 100 Hz. This is consistent with the fact that the first natural frequency of this structure -- the most likely to dominate the response -- is 100.8 Hz. The second plot (Figure 18.3) shows the applied load (impulse load) acting at GRID 105 as a function of time.

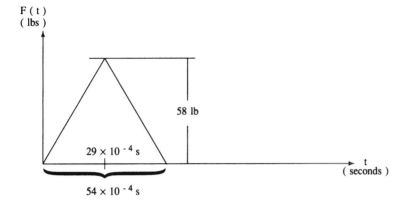

Figure 18.1

```
N A S T R A N   E X E C U T I V E   C O N T R O L   D E C K   E C H O


   ID    AAAAAA,AAAABBBB
   TIME  5
   SOL   27    $  SOL=27 CORRESPONDS TO TRANSIENT RESPONSE
   CEND
```

Exhibit 18.1   Executive Control Deck

```
                  C A S E    C O N T R O L   D E C K   E C H O
CARD
COUNT
  1      TITLE= TRANSIENT RESPONSE
  2      $$
  3      DLOAD=  100  $ TO SELECT THE LOAD ACTING AT NODE 105
  4      $$
  5      TSTEP=200     $ TO SELECT THE TSTEP  CARD
  6      SET  1000=   102
  7      DISP= 1000
  8      ECHO=BOTH
  9      $$$$$$$$$$$$$  P L O T S $$$$$$$$$$$$$$$$$$$$$
 10      $$
 11      OUTPUT(XYPLOT)
 12      XTAXIS=YES
 13      XBAXIS=YES
 14      XTGRID  LINES=  YES
 15      XBGRID  LINES = YES
 16      YTGRID LINES = YES
 17      YBGRID LINES = YES
 18      $$
 19      XTITLE=   T I M E   IN   S E C O N D S
 20      YTITLE=  NODE  102   DISPLACEMENT
 21      XYPLOT  DISP   /102(T2)
 22      $$
 23      YTITLE=  EXTERNAL LOAD (EXCITATION)  AT 105
 24      XYPLOT  OLOAD  /105(T2)
 25      BEGIN BULK
```

Exhibit 18.2   Case Control Deck

```
              I N P U T   B U L K   D A T A   D E C K   E C H O
     .   1  ..  2  ..  3  ..  4  ..  5  ..  6  ..  7  ..  8  ..  9  ..  10  .
$$
$$$$   DYNAMIC   LOAD
$$
TLOAD1  100    101             0      104
DAREA   101    105     2       58.00
TABLED1 104                                                          +NNNN
+NNNN   0.0    0.0     6.0-4   0.0    30.-4  1.000  60.00-4 .00       +M
+M      999999. .0     ENDT
$$
$$   DASHPOTS   (DAMPERS)
$$
CDAMP2  4455   1.5     101     2
CDAMP2  3322   1.5     109     2
$$
$$
$$
$$   NUMERICAL  INTEGRATION PARAMETERS
$$
TSTEP   200    130     3.-4    2
$$
$$
```

Exhibit 18.3 Input Bulk Data Deck

```
GRID     101            00.00  0.00  0.00  0.00  1345
GRID     102            5.     0.00  0.00  0.00  1345
GRID     103            10.00  0.00  0.00  0.00  1345
GRID     104            15.    0.00  0.00  0.00  1345
GRID     105            20.00  0.00  0.00  0.00  1345
GRID     106            25.00  0.00  0.00  0.00  1345
GRID     107            30.00  0.00  0.00  0.00  1345
GRID     108            35.00  0.00  0.00  0.00  1345
GRID     109            40.    0.00  0.00  0.00  1345
CBAR     1001 5555      101    102   1.000 0.000 1.000
CBAR     1002 5555      102    103   1.000 0.000 1.000
CBAR     1003 5555      103    104   1.000 0.000 1.000
CBAR     1004 5555      104    105   1.000 0.000 1.000
CBAR     1005 5555      105    106   1.000 0.000 1.000
CBAR     1006 5555      106    107   1.000 0.000 1.000
CBAR     1007 5555      107    108   1.000 0.000 1.000
CBAR     1008 5555      108    109   1.000 0.000 1.000
PBAR     5555 8888      4.00          1.333
MAT1     8888 30.+6     .29           7.0-4
$$ CONCENTRATED MASS AT POINT 102
CMASS2   123456 .02     102    2
$$
$$ SPRINGS AT NODE 101 AND 109
CELAS2   2233 10.+4     101    2
CELAS2   2244 10.+4     109    2
ENDDATA
```

Exhibit 18.3 (continued) Input Bulk Data Deck

S O R T E D   B U L K   D A T A   E C H O

CARD COUNT	. 1 .	. 2 .	. 3 .	. 4 .	. 5 .	. 6 .	. 7 .	. 8 .	. 9 .	. 10 .
1-	CBAR	1001	5555	101	102	1.000	0.000	1.000		
2-	CBAR	1002	5555	102	103	1.000	0.000	1.000		
3-	CBAR	1003	5555	103	104	1.000	0.000	1.000		
4-	CBAR	1004	5555	104	105	1.000	0.000	1.000		
5-	CBAR	1005	5555	105	106	1.000	0.000	1.000		
6-	CBAR	1006	5555	106	107	1.000	0.000	1.000		
7-	CBAR	1007	5555	107	108	1.000	0.000	1.000		
8-	CBAR	1008	5555	108	109	1.000	0.000	1.000		
9-	CDAMP2	3322	1.5	109	2					
10-	CDAMP2	4455	1.5	101	2					
11-	CELAS2	2233	10.+4	101	2					
12-	CELAS2	2244	10.+4	109	2					
13-	CMASS2	123456	.02	102	2					
14-	DAREA	101	105	2	58.00					
15-	GRID	101		00.00	0.00	0.00		1345		
16-	GRID	102		5.	0.00	0.00		1345		
17-	GRID	103		10.00	0.00	0.00		1345		
18-	GRID	104		15.	0.00	0.00		1345		
19-	GRID	105		20.00	0.00	0.00		1345		
20-	GRID	106		25.00	0.00	0.00		1345		
21-	GRID	107		30.00	0.00	0.00		1345		
22-	GRID	108		35.00	0.00	0.00		1345		
23-	GRID	109		40.	0.00	0.00		1345		

Exhibit 18.4   Bulk Data Deck sorted by alphabetical order

```
24-  MAT1     8888    30.0+6            .29     7.0-4
25-  PBAR     5555    8888     4.00             1.333
26-  TABLED1  104
27-  +NNNN    0.0     0.0      6.0-4    0.0     30.-4    1.000   60.00-4  .00    +NNNN
28-  +M       999999. .0       ENDT                                             +M
29-  TLOAD1   100     101               0       104
30-  TSTEP    200     130      3.-4     2
     ENDDATA
```

Exhibit 18.4 (continued)   Bulk Data Deck sorted by alphabetical order

POINT-ID =   102

			DISPLACEMENT VECTOR				
TIME	TYPE	T1	T2	T3	R1	R2	R3
0.0		0.0	0.0	0.0	0.0	0.0	0.0
6.000000E-04	G	0.0	0.0	0.0	0.0	0.0	0.0
1.200000E-03	G	0.0	6.523056E-07	0.0	0.0	0.0	1.243290E-06
1.800000E-03	G	0.0	1.753679E-05	0.0	0.0	0.0	9.780321E-06
2.400000E-03	G	0.0	9.321395E-05	0.0	0.0	0.0	2.676819E-05
3.000000E-03	G	0.0	2.689907E-04	0.0	0.0	0.0	5.215155E-05
3.600000E-03	G	0.0	5.534931E-04	0.0	0.0	0.0	8.715841E-05
4.200000E-03	G	0.0	9.080537E-04	0.0	0.0	0.0	1.223745E-04
4.800000E-03	G	0.0	1.240797E-03	0.0	0.0	0.0	1.509431E-04
5.400000E-03	G	0.0	1.443906E-03	0.0	0.0	0.0	1.679632E-04

Exhibit 18.5   Displacement vector as a function of time

POINT-ID =   102

TIME	TYPE	T1	DISPLACEMENT VECTOR T2	T3	R1	R2	R3
6.000000E-03	G	0.0	1.446824E-03	0.0	0.0	0.0	1.656685E-04
6.599999E-03	G	0.0	1.230023E-03	0.0	0.0	0.0	1.396333E-04
7.199999E-03	G	0.0	8.212753E-04	0.0	0.0	0.0	9.617433E-05
7.799999E-03	G	0.0	3.021141E-04	0.0	0.0	0.0	4.064719E-05
8.400000E-03	G	0.0	-2.274313E-04	0.0	0.0	0.0	-2.321800E-05
9.000001E-03	G	0.0	-7.043010E-04	0.0	0.0	0.0	-8.581433E-05
9.600001E-03	G	0.0	-1.098000E-03	0.0	0.0	0.0	-1.339571E-04
1.020000E-02	G	0.0	-1.364713E-03	0.0	0.0	0.0	-1.611113E-04
1.080000E-02	G	0.0	-1.446238E-03	0.0	0.0	0.0	-1.669130E-04
1.140000E-02	G	0.0	-1.313810E-03	0.0	0.0	0.0	-1.506146E-04
1.200000E-02	G	0.0	-9.829644E-04	0.0	0.0	0.0	-1.114943E-04
1.260001E-02	G	0.0	-4.996868E-04	0.0	0.0	0.0	-5.610023E-05
1.320001E-02	G	0.0	6.195448E-05	0.0	0.0	0.0	5.269219E-06
1.380001E-02	G	0.0	6.000670E-04	0.0	0.0	0.0	6.521626E-05
1.440001E-02	G	0.0	1.024734E-03	0.0	0.0	0.0	1.185004E-04
1.500001E-02	G	0.0	1.296867E-03	0.0	0.0	0.0	1.558925E-04
1.560001E-02	G	0.0	1.408699E-03	0.0	0.0	0.0	1.690103E-04
1.620001E-02	G	0.0	1.344981E-03	0.0	0.0	0.0	1.567758E-04
1.680001E-02	G	0.0	1.091234E-03	0.0	0.0	0.0	1.241810E-04
1.740001E-02	G	0.0	6.685082E-04	0.0	0.0	0.0	7.480910E-05
1.800001E-02	G	0.0	1.386812E-04	0.0	0.0	0.0	1.383067E-05
1.860001E-02	G	0.0	-4.194753E-04	0.0	0.0	0.0	-4.944365E-05

Exhibit 18.5 (continued)   Displacement vector as a function of time

POINT-ID =   102

TIME	TYPE	T1	DISPLACEMENT VECTOR T2	T3	R1	R2	R3
1.920000E-02	G	0.0	-9.163388E-04	0.0	0.0	0.0	-1.038861E-04
1.980000E-02	G	0.0	-1.264880E-03	0.0	0.0	0.0	-1.439507E-04
2.040000E-02	G	0.0	-1.413483E-03	0.0	0.0	0.0	-1.655619E-04
2.100000E-02	G	0.0	-1.366052E-03	0.0	0.0	0.0	-1.641187E-04
2.160000E-02	G	0.0	-1.150968E-03	0.0	0.0	0.0	-1.370961E-04
2.220000E-02	G	0.0	-7.899165E-04	0.0	0.0	0.0	-9.027994E-05
2.280000E-02	G	0.0	-3.094483E-04	0.0	0.0	0.0	-3.216368E-05
2.340000E-02	G	0.0	2.313094E-04	0.0	0.0	0.0	2.969679E-05
2.400000E-02	G	0.0	7.498759E-04	0.0	0.0	0.0	8.860909E-05
2.460000E-02	G	0.0	1.164165E-03	0.0	0.0	0.0	1.346439E-04
2.519999E-02	G	0.0	1.405516E-03	0.0	0.0	0.0	1.607192E-04
2.579999E-02	G	0.0	1.429342E-03	0.0	0.0	0.0	1.643437E-04
2.639999E-02	G	0.0	1.239862E-03	0.0	0.0	0.0	1.466325E-04
2.699999E-02	G	0.0	8.869740E-04	0.0	0.0	0.0	1.074892E-04
2.759999E-02	G	0.0	4.332889E-04	0.0	0.0	0.0	5.158222E-05
2.819999E-02	G	0.0	-7.393249E-05	0.0	0.0	0.0	-1.189250E-05
2.879999E-02	G	0.0	-5.837897E-04	0.0	0.0	0.0	-7.185272E-05
2.939999E-02	G	0.0	-1.027621E-03	0.0	0.0	0.0	-1.216378E-04
2.999999E-02	G	0.0	-1.331724E-03	0.0	0.0	0.0	-1.548455E-04
3.059999E-02	G	0.0	-1.443413E-03	0.0	0.0	0.0	-1.662157E-04
3.119998E-02	G	0.0	-1.338851E-03	0.0	0.0	0.0	-1.530697E-04

Exhibit 18.5 (continued)   Displacement vector as a function of time

POINT-ID =   102

| TIME | TYPE | T1 | D I S P L A C E M E N T   V E C T O R | | | | |
			T2	T3	R1	R2	R3
3.179999E-02	G	0.0	-1.030620E-03	0.0	0.0	0.0	-1.193029E-04
3.239999E-02	G	0.0	-5.741073E-04	0.0	0.0	0.0	-6.949900E-05
3.299999E-02	G	0.0	-5.411862E-05	0.0	0.0	0.0	-9.398742E-06
3.360000E-02	G	0.0	4.545743E-04	0.0	0.0	0.0	5.374961E-05
3.420000E-02	G	0.0	8.988613E-04	0.0	0.0	0.0	1.085424E-04
3.480000E-02	G	0.0	1.233349E-03	0.0	0.0	0.0	1.469395E-04
3.540000E-02	G	0.0	1.406471E-03	0.0	0.0	0.0	1.642313E-04
3.600001E-02	G	0.0	1.381241E-03	0.0	0.0	0.0	1.594149E-04
3.660001E-02	G	0.0	1.151598E-03	0.0	0.0	0.0	1.315240E-04
3.720001E-02	G	0.0	7.472575E-04	0.0	0.0	0.0	8.494606E-05
3.780001E-02	G	0.0	2.283167E-04	0.0	0.0	0.0	2.682191E-05
3.840002E-02	G	0.0	-3.171253E-04	0.0	0.0	0.0	-3.415413E-05
3.900002E-02	G	0.0	-7.995992E-04	0.0	0.0	0.0	-9.151603E-05

Exhibit 18.5 (continued)   Displacement vector as a function of time

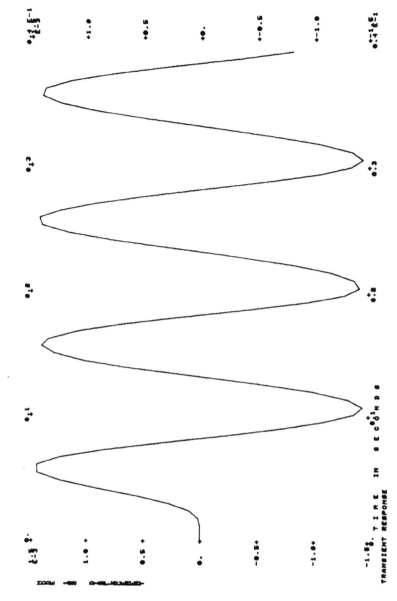

Figure 18.2 Displacement at Grid 102 in the Y-direction as a function of time.

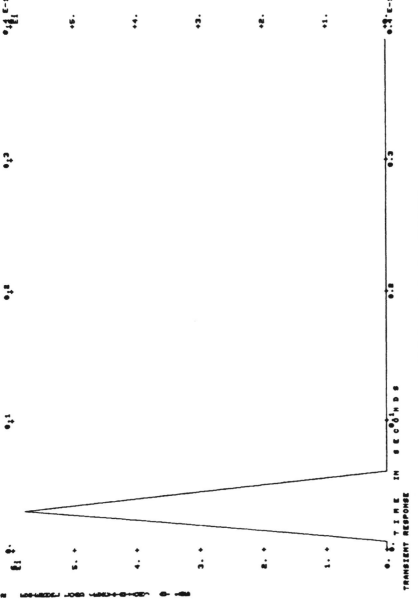

Figure 18.3 Impulse load at Grid 105.

19

Problem 19

19.1 Statement of the problem

Consider again the beam structure of Problem 14 (Figure 14.1) and assume that two dashpots ($c = 1.5$ lb-s/in) have been added at the ends, as in Problem 18. Determine the transient response of this beam when it vibrates freely under the action of gravity assuming the initial configuration described in Figure 19.1. No external loads other than the effect of gravity are considered.

19.2 Cards introduced

Case Control Deck

ACCELERATION
IC
LOADSET
VELOCITY

Bulk Data Deck

LSEQ
TIC

19.3 MSC/NASTRAN formulation

The same model of Problem 18 is used. This problem demonstrates the use of static loads (in this case, gravity) in dynamic analyses. It also explains how to specify non-homogeneous (different than zero) initial conditions for transient analysis problems. The analysis is performed using direct transient analysis, SOL 27.

19.4 Input Data Deck

Exhibit 19.1 shows the Executive Control Deck and Exhibit 19.2 shows the Case Control Deck.

Examine the Case Control Deck. The TSTEP card selects the TSTEP card in the Bulk Data Deck. Recall that this card is always required in SOL 27. The velocity and the acceleration, as a function of time, will be printed for GRID 102. In addition, the element forces will be printed for element 1001.

Static loads are incorporated into MSC/NASTRAN dynamic analysis following an odd procedure that, if understood correctly, should pose no difficulty.[1] This procedure will be explained in detail, without offering a logical rationale. At this moment, we will just say that a DLOAD card and a LOADSET card are required in the Case Control Deck.

The IC=90 card selects from the Bulk Data Deck the TIC cards required to determine the initial conditions. Recall that in Problem 18 no specifications regarding initial conditions were made; thus, the program assumed that at $t=0$ both velocities and displacements were zero for all GRID points. The first part of the plot subdeck (cards 24 to 30) is similar to that of Problem 18. The second part of this subdeck requests two plots: one to show the velocity at GRID 102 as a function of time, and a second to show the acceleration at GRID 102 as a function of time. Exhibits 19.3 and 19.4 show the Bulk Data Deck. The TIC cards included in the Bulk Data Deck specify the initial conditions for the grids of the model, in this case initial displacements according to Figure 19.1.

The procedure to incorporate static loads in dynamic analyses is as follows. The DLOAD=50011 card in the Case Control Deck points to a TLOAD1 card. This TLOAD1 card serves two purposes. First, it points to a TABLED1 card (ID=666) that defines a function of time equal to 1 (constant with time). Second, it points to a LSEQ card through the 5002 in the third field. In other words, the ID=5002 that goes in the third field of the TLOAD1 card, points to the 5002 that goes in the third field of the LSEQ card. The LSEQ card has an ID number of 5000, which is consistent with LOADSET=5000 in the Case Control Deck. Finally, the LSEQ card points to the GRAV card via the ID number (25) specified in the fourth field.

The result of this operation is that the GRAV card is incorporated into the dynamic analysis. The TABLED1 card defines the "amplitude" of the load as a function of time -- in this example it is a function equal to 1.0 for any value of t in the range of interest -- since the gravity acceleration is constant with time.

---

[1] Static loads refer to loads that are typical of SOL 24 runs; namely, loads defined by the following cards: GRAV, FORCE, MOMENT, PLOAD1, etc. These loads are not time dependent *per se*.

This procedure is necessary because static loads cannot be incorporated in a dynamic analysis simply by using a DLOAD card in the Case Control Deck. For example, DLOAD = 50011 in the Case Control Deck accompanied by a GRAV card with ID = 50011 in the Bulk Data Deck would have been unacceptable. Other types of static load cards can be incorporated into dynamic analyses using an analogous procedure. This subject is discussed in more detail in Appendix V.

Finally, notice that the TSTEP card selects a time step about 1/10th the period of the first mode ($T_1 = .0099$ s, from Problem 14, Exhibit 14.6). This is adequate to get a good overall picture of the response of the structure. The number of integration steps is 1300.

19.5 Results

Exhibit 19.5 shows the velocity at GRID 102 as a function of time for a few initial and final time steps to avoid printing a voluminous table. Exhibit 19.6 shows the value of the acceleration at GRID 102 as a function of time. Note that the velocity and acceleration decrease dramatically after 1300 steps. The same phenomenon is observed for the bending moment (plane 2) and shear in element 1001 (see Exhibit 19.7). The structure eventually dissipates the initial potential energy and since no external forces other than gravity are acting, its configuration approaches the so-called static response (in this case, the response of the structure to a gravity load, applied statically).

The plots (Figures 19.2 and 19.3) show the acceleration and velocity at GRID 102 as a function of time. It can be seen that they converge to zero, as expected, for large values of t.

19.6 Additional comments

If a static analysis (SOL 24) of this structure were performed, the results would be similar to those obtained after 1300 integration steps in transient analysis.

Exhibit 19.8 includes the output file of such run. As expected, the values for moment and shear given at the bottom of this Exhibit are very close to the values shown at the bottom of Exhibit 19.7. This demonstrates that for a sufficiently large time t the structure approaches its static response.

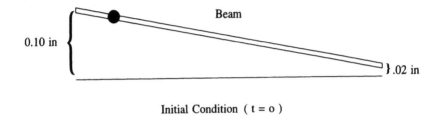

0.10 in

Beam

} .02 in

Initial Condition (t = o)

Figure 19.1

```
N A S T R A N   E X E C U T I V E   C O N T R O L   D E C K   E C H O

     ID   A,B
     TIME  5
     SOL    27
     CEND
```

Exhibit 19.1   Executive Control Deck

```
                    C A S E    C O N T R O L    D E C K    E C H O
CARD
COUNT
   1      TITLE=    TRANSIENT RESPONSE
   2      SUBTITLE= INITIAL  CONDITIONS PLUS GRAVITY
   3      $$
   4      TSTEP=200    $ TO SELECT THE TSTEP   CARD
   5      $$
   6      SET  1000=  102
   7      VELOCITY= 1000
   8      ACCELERATION  = 1000
   9      SET 55=  1001
  10      ELFORCES=55
  11      $$
  12      $$              LET US INCLUDE THE EFFECT OF GRAVITY
  13      $$
  14      DLOAD=  50011 $  SELECTS TLOAD   CARD
  15      LOADSET=5000  $  SELECTS LSEQ   CARD
  16      $$
  17      $$ INITIAL   CONDITIONS
  19      IC=90
  20      $$
  21      ECHO=BOTH
  22      $************    PLOTS   ***************
  24      OUTPUT(XYPLOT)
  25      XTAXIS=YES
  26      XBAXIS=YES
  27      XTGRID  LINES=  YES
  28      XBGRID  LINES = YES
  29      YTGRID LINES = YES
  30      YBGRID LINES = YES
  31      $$
  32      XTITLE=   T I M E   IN   S E C O N D S
  33      YTITLE=  NODE  102  V E L O C I T Y
  34      XYPLOT  VELO   /102(T2)
  35      $$
  36      YTITLE= NODE  102  A C C E L.
  37      XYPLOT  ACCE   /102(T2)
  38      $$
  39      BEGIN BULK
```

Exhibit 19.2 Case Control Deck

```
                    I N P U T   B U L K   D A T A   D E C K   E C H O
.     1 ..   2 ..   3 ..   4 ..   5 ..   6 ..   7 ..   8 ..   9 ..   10 .
$$
$$ INITIAL CONDITIONS
$$
TIC    90    101    2    .10
TIC    90    102    2    .09
TIC    90    103    2    .08
TIC    90    104    2    .07
TIC    90    105    2    .06
TIC    90    106    2    .05
TIC    90    107    2    .04
TIC    90    108    2    .03
TIC    90    109    2    .02
$$
$$ TO INCLUDE THE GRAVITY "LOAD"
$$
TLOAD1  50011  5002                666
$$
TABLED1 666
+PLO   -3.    1.000   10.+9   1.00   ENDT                        +PLO
$$
LSEQ   5000   5002    25
$$
GRAV   25     386.    .0     -1.    .0
```

Exhibit 19.3 Input Bulk Data Deck

```
                I N P U T   B U L K   D A T A   D E C K   E C H O
    .  1  ..  2  ..  3  ..  4  ..  5  ..  6  ..  7  ..  8  ..  9  ..  10  .

$$
$$ DASHPOTS (DAMPERS)
$$
CDAMP2   4455   25.     101    2
CDAMP2   3322   25.     109    2
$$$$$$$$$$$$$$$$$$$$$$$$$$$$$$$$$$
$$ NUMERICAL INTEGRATION PARAMETERS
$$$$$$$$$$$$$$$$$$$$$$$$$$$$$$$$$$
TSTEP    200    1300    1.-3   2
GRID     101    00.00   0.00   0.00   1345
GRID     102    5.      0.00   0.00   1345
GRID     103    10.00   0.00   0.00   1345
GRID     104    15.     0.00   0.00   1345
GRID     105    20.00   0.00   0.00   1345
GRID     106    25.00   0.00   0.00   1345
GRID     107    30.00   0.00   0.00   1345
GRID     108    35.00   0.00   0.00   1345
GRID     109    40.     0.00   0.00   1345
```

Exhibit 19.3 (continued) Input Bulk Data Deck

```
              I N P U T   B U L K   D A T A   D E C K   E C H O

    1  ..   2  ..   3  ..   4  ..   5  ..   6  ..   7  ..   8  ..   9  ..  10  .
CBAR     1001 5555   101    102    1.000  0.000  1.000
CBAR     1002 5555   102    103    1.000  0.000  1.000
CBAR     1003 5555   103    104    1.000  0.000  1.000
CBAR     1004 5555   104    105    1.000  0.000  1.000
CBAR     1005 5555   105    106    1.000  0.000  1.000
CBAR     1006 5555   106    107    1.000  0.000  1.000
CBAR     1007 5555   107    108    1.000  0.000  1.000
CBAR     1008 5555   108    109    1.000  0.000  1.000
PBAR     5555 8888   4.00          1.333
MAT1     8888 30.0+6 .29    7.0-4
$$
$$ CONCENTRATED MASS  AT POINT 102
$$
CMASS2 123456 .02    102    2
$$
$$ SPRINGS AT NODE 101  AND  109
$$
CELAS2 2233  10.+4   101    2
CELAS2 2244  10.+4   109    2
ENDDATA
```

Exhibit 19.3 (continued) Input Bulk Data Deck

```
                              S O R T E D   B U L K   D A T A   E C H O

CARD
COUNT     .   1  ..  2    ..  3    ..  4   ..  5   ..  6   ..  7   ..  8  ..  9  .. 10  .
  1-       CBAR   1001    5555     101     102     1.000   0.000   1.000
  2-       CBAR   1002    5555     102     103     1.000   0.000   1.000
  3-       CBAR   1003    5555     103     104     1.000   0.000   1.000
  4-       CBAR   1004    5555     104     105     1.000   0.000   1.000
  5-       CBAR   1005    5555     105     106     1.000   0.000   1.000
  6-       CBAR   1006    5555     106     107     1.000   0.000   1.000
  7-       CBAR   1007    5555     107     108     1.000   0.000   1.000
  8-       CBAR   1008    5555     108     109     1.000   0.000   1.000
  9-       CDAMP2 3322    25.      109     2
 10-       CDAMP2 4455    25.      101     2
 11-       CELAS2 2233    10.+4    101     2
 12-       CELAS2 2244    10.+4    109     2
 13-       CMASS2 123456  .02      102     2
 14-       GRAV   25              386.     .0      -1.     .0
 15-       GRID   101             00.00    0.00    0.00            1345
 16-       GRID   102             5.       0.00    0.00            1345
 17-       GRID   103             10.00    0.00    0.00            1345
 18-       GRID   104             15.      0.00    0.00            1345
 19-       GRID   105             20.00    0.00    0.00            1345
 20-       GRID   106             25.00    0.00    0.00            1345
 21-       GRID   107             30.00    0.00    0.00            1345
 22-       GRID   108             35.00    0.00    0.00            1345
 23-       GRID   109             40.      0.00    0.00            1345
```

Exhibit 19.4   Bulk Data Deck sorted by alphabetical order

SORTED BULK DATA ECHO

CARD COUNT	. 1 ..	2 ..	3 ..	4 ..	5 ..	6 ..	7 ..	8 ..	9 ..	10 .
24-	LSEQ	5000	5002	25						
25-	MAT1	8888	30.0+6		.29	7.0-4				
26-	PBAR	5555	8888	4.00		1.333				
27-	TABLED1	666								+PLO
28-	+PLO	-3.	1.000	10.+9	1.00	ENDT				
29-	TIC	90	101	2	.10					
30-	TIC	90	102	2	.09					
31-	TIC	90	103	2	.08					
32-	TIC	90	104	2	.07					
33-	TIC	90	105	2	.06					
34-	TIC	90	106	2	.05					
35-	TIC	90	107	2	.04					
36-	TIC	90	108	2	.03					
37-	TIC	90	109	2	.02					
38-	TLOAD1	50011	5002			666				
39-	TSTEP	200	1300	1.-3	2					
	ENDDATA									

Exhibit 19.4 (continued) Bulk Data Deck sorted by alphabetical order

POINT-ID =   102

| TIME | TYPE | VELOCITY VECTOR | | | | | |
		T1	T2	T3	R1	R2	R3
0.0	G	0.0	-1.270698E+01	0.0	0.0	0.0	9.819679E-01
2.000000E-03	G	0.0	-3.789610E+01	0.0	0.0	0.0	3.874042E-02
4.000000E-03	G	0.0	-4.775321E+00	0.0	0.0	0.0	-3.200610E+00
6.000001E-03	G	0.0	-7.291243E+00	0.0	0.0	0.0	3.726795E-01
8.000000E-03	G	0.0	2.148543E+01	0.0	0.0	0.0	2.627390E+00
1.000000E-02	G	0.0	1.563787E+01	0.0	0.0	0.0	1.083533E+00
1.200000E-02	G	0.0	-1.432896E+01	0.0	0.0	0.0	-1.028876E+00
1.400000E-02	G	0.0	-1.512595E+01	0.0	0.0	0.0	-2.439057E+00
1.600000E-02	G	0.0	-7.241820E+00	0.0	0.0	0.0	-2.307876E-01
1.800000E-02	G	0.0	1.790979E+01	0.0	0.0	0.0	1.940551E+00
2.000000E-02	G	0.0	1.430603E+01	0.0	0.0	0.0	1.631643E+00
2.200000E-02	G	0.0	-5.511416E+00	0.0	0.0	0.0	-6.851298E-01
2.400000E-02	G	0.0	-1.706560E+01	0.0	0.0	0.0	-2.166020E+00
SOME LINES HAVE BEEN DELETED							
1.284004E+00	G	0.0	2.370888E-04	0.0	0.0	0.0	-1.439117E-04
1.286004E+00	G	0.0	3.728589E-04	0.0	0.0	0.0	-3.313182E-05
1.288004E+00	G	0.0	-5.556418E-04	0.0	0.0	0.0	1.889098E-04
1.290004E+00	G	0.0	2.177315E-04	0.0	0.0	0.0	-1.679501E-04
1.292004E+00	G	0.0	2.531419E-04	0.0	0.0	0.0	-2.020478E-05
1.294004E+00	G	0.0	-5.374542E-04	0.0	0.0	0.0	1.848950E-04
1.296005E+00	G	0.0	3.493896E-04	0.0	0.0	0.0	-1.718751E-04
1.298005E+00	G	0.0	2.401158E-04	0.0	0.0	0.0	5.106943E-06
1.300005E+00	G	0.0	-5.856620E-04	0.0	0.0	0.0	1.692515E-04

Exhibit 19.5   Velocity vector as a function of time

POINT-ID =   102

TIME	TYPE	T1	ACCELERATION T2	VECTOR T3	R1	R2	R3
0.0	G	0.0	-2.541396E+04	0.0	0.0	0.0	1.963936E+03
2.000000E-03	G	0.0	3.176926E+03	0.0	0.0	0.0	-2.759136E+03
4.000000E-03	G	0.0	6.787352E+03	0.0	0.0	0.0	2.120391E+03
6.000001E-03	G	0.0	4.224377E+03	0.0	0.0	0.0	7.443082E+02
8.000000E-03	G	0.0	1.501727E+04	0.0	0.0	0.0	6.760352E+02
1.000000E-02	G	0.0	-1.899050E+04	0.0	0.0	0.0	-1.260898E+03
1.200000E-02	G	0.0	-5.441993E+03	0.0	0.0	0.0	-1.214640E+03
1.400000E-02	G	0.0	2.570685E+02	0.0	0.0	0.0	2.251770E+02
1.600000E-02	G	0.0	1.113646E+04	0.0	0.0	0.0	1.618914E+03
1.800000E-02	G	0.0	8.496388E+03	0.0	0.0	0.0	4.506071E+02
2.000000E-02	G	0.0	-1.012635E+04	0.0	0.0	0.0	-8.022775E+02
2.200000E-02	G	0.0	-8.097006E+03	0.0	0.0	0.0	-1.278017E+03

SOME LINES HAVE BEEN DELETED

TIME	TYPE	T1	ACCELERATION T2	VECTOR T3	R1	R2	R3
1.284004E+00	G	0.0	-1.737202E+00	0.0	0.0	0.0	4.670144E-01
1.286004E+00	G	0.0	1.515356E+00	0.0	0.0	0.0	-6.760623E-01
1.288004E+00	G	0.0	1.143261E-01	0.0	0.0	0.0	2.468007E-01
1.290004E+00	G	0.0	-1.685775E+00	0.0	0.0	0.0	4.122124E-01
1.292004E+00	G	0.0	1.653834E+00	0.0	0.0	0.0	-6.886987E-01
1.294004E+00	G	0.0	-3.036734E-02	0.0	0.0	0.0	3.219193E-01
1.296005E+00	G	0.0	-1.580755E+00	0.0	0.0	0.0	3.557480E-01
1.298005E+00	G	0.0	1.699101E+00	0.0	0.0	0.0	-7.034278E-01
1.300005E+00	G	0.0	-2.616672E-01	0.0	0.0	0.0	3.845499E-01

Exhibit 19.6   Acceleration vector as a function of time

ELEMENT-ID = 1001

FORCES IN BAR ELEMENTS (CBAR)

TIME	BEND-MOMENT-END-A		BEND-MOMENT-END-B		SHEAR		FORCE	TORQUE
	PLANE 1	PLANE 2	PLANE 1	PLANE 2	PLANE 1	PLANE 2		
0.0	0.0	9.597598E+04	0.0	-9.597598E+04	0.0	3.839039E+04	0.0	0.0
2.000000E-03	0.0	1.455192E-11	0.0	1.079095E+03	0.0	-2.158190E+02	0.0	0.0
4.000000E-03	0.0	7.275958E-12	0.0	-4.272083E+03	0.0	8.544167E+02	0.0	0.0
6.000001E-03	0.0	-2.182787E-11	0.0	-4.140827E+03	0.0	8.281654E+02	0.0	0.0
8.000000E-03	0.0	2.182787E-11	0.0	-4.857214E+03	0.0	9.714429E+02	0.0	0.0
1.000000E-02	0.0	2.910383E-11	0.0	7.300106E+03	0.0	-1.460021E+03	0.0	0.0
1.200000E-02	0.0	-4.729372E-11	0.0	3.770939E+03	0.0	-7.541879E+02	0.0	0.0
1.400000E-02	0.0	2.091838E-11	0.0	-3.574542E+01	0.0	7.149084E+00	0.0	0.0
1.600000E-02	0.0	2.910383E-11	0.0	-7.173418E+03	0.0	1.434684E+03	0.0	0.0
1.800000E-02	0.0	-5.093170E-11	0.0	-3.253916E+03	0.0	6.507832E+02	0.0	0.0
2.000000E-02	0.0	2.182787E-11	0.0	4.655198E+03	0.0	-9.310397E+02	0.0	0.0
SOME LINES HAVE BEEN DELETED								
1.284004E+00	0.0	-5.809397E-11	0.0	-1.187432E+02	0.0	2.374864E+01	0.0	0.0
1.286004E+00	0.0	4.638423E-11	0.0	-1.363934E+02	0.0	2.727868E+01	0.0	0.0
1.288004E+00	0.0	1.205080E-11	0.0	-1.294068E+02	0.0	2.588135E+01	0.0	0.0
1.290004E+00	0.0	-5.911716E-11	0.0	-1.191230E+02	0.0	2.382459E+01	0.0	0.0
1.292004E+00	0.0	4.672529E-11	0.0	-1.370709E+02	0.0	2.741418E+01	0.0	0.0
1.294004E+00	0.0	1.250555E-11	0.0	-1.283389E+02	0.0	2.566779E+01	0.0	0.0
1.296005E+00	0.0	-5.923084E-11	0.0	-1.196228E+02	0.0	2.392457E+01	0.0	0.0
1.298005E+00	0.0	4.672529E-11	0.0	-1.376140E+02	0.0	2.752279E+01	0.0	0.0
1.300005E+00	0.0	1.273293E-11	0.0	-1.272233E+02	0.0	2.544466E+01	0.0	0.0

Exhibit 19.7 Forces in BAR elements as a function of time

```
N A S T R A N   E X E C U T I V E   C O N T R O L   D E C K   E C H O

    ID   A,B
    TIME  5
    SOL  24
    CEND
                        C A S E    C O N T R O L   D E C K   E C H O
CARD
COUNT
  1   TITLE=   S T A T I C    A N A L Y S I S
  2   SUBTITLE= OF THE BEAM-SPRING SYSTEM ANALYZED IN THE PREVIOUS RUN
  3   $$
  4   LOAD  25   $ TO SELECT GRAVITY
  5   SET 55=  1001
  6   ELFORCES=55
  7   $$
  8   ECHO=BOTH
  9   BEGIN BULK
              I N P U T   B U L K   D A T A   D E C K   E C H O

.   1 ..  2 ..  3 ..  4 ..  5 ..  6 ..  7 ..  8 ..  9
$$
$$
$$  TO INCLUDE THE GRAVITY "LOAD"
$$
GRAV    25           386.    .0     -1.     .0
$$
$$
$$$$$$$$$$$$$$$$$$$$$$$$$$$$$$$$$$$$$
GRID       101        00.00   0.00   0.00      1345
GRID       102        5.      0.00   0.00      1345
GRID       103        10.00   0.00   0.00      1345
GRID       104        15.     0.00   0.00      1345
GRID       105        20.00   0.00   0.00      1345
GRID       106        25.00   0.00   0.00      1345
GRID       107        30.00   0.00   0.00      1345
GRID       108        35.00   0.00   0.00      1345
GRID       109        40.     0.00   0.00      1345
```

Exhibit 19.8   Static analysis

```
CBAR        1001 5555         101    102   1.000   0.000   1.000
CBAR        1002 5555         102    103   1.000   0.000   1.000
CBAR        1003 5555         103    104   1.000   0.000   1.000
CBAR        1004 5555         104    105   1.000   0.000   1.000
CBAR        1005 5555         105    106   1.000   0.000   1.000
CBAR        1006 5555         106    107   1.000   0.000   1.000
CBAR        1007 5555         107    108   1.000   0.000   1.000
CBAR        1008 5555         108    109   1.000   0.000   1.000
PBAR    5555    8888    4.00            1.333
MAT1    8888    30.0+6          .29    7.0-4
$$  CONCENTRATED MASS    AT POINT   102
CMASS2 123456  .02      102     2
$$ SPRINGS AT NODE   101     AND   109
CELAS2 2233    10.+4    101     2
CELAS2 2244    10.+4    109     2
ENDDATA
                        S O R T E D   B U L K   D A T A   E C H O
CARD
COUNT   .  1  ..  2  ..  3  ..  4  ..  5  ..  6  ..  7  ..  8
  1-    CBAR    1001    5555    101     102     1.000   0.000   1.000
  2-    CBAR    1002    5555    102     103     1.000   0.000   1.000
  3-    CBAR    1003    5555    103     104     1.000   0.000   1.000
  4-    CBAR    1004    5555    104     105     1.000   0.000   1.000
  5-    CBAR    1005    5555    105     106     1.000   0.000   1.000
  6-    CBAR    1006    5555    106     107     1.000   0.000   1.000
  7-    CBAR    1007    5555    107     108     1.000   0.000   1.000
  8-    CBAR    1008    5555    108     109     1.000   0.000   1.000
  9-    CELAS2  2233    10.+4   101     2
 10-    CELAS2  2244    10.+4   109     2
 11-    CMASS2  123456  .02     102     2
 12-    GRAV    25              386.    .0      -1.     .0
 13-    GRID    101             00.00   0.00    0.00            1345
 14-    GRID    102             5.      0.00    0.00            1345
 15-    GRID    103             10.00   0.00    0.00            1345
 16-    GRID    104             15.     0.00    0.00            1345
 17-    GRID    105             20.00   0.00    0.00            1345
 18-    GRID    106             25.00   0.00    0.00            1345
 19-    GRID    107             30.00   0.00    0.00            1345
 20-    GRID    108             35.00   0.00    0.00            1345
 21-    GRID    109             40.     0.00    0.00            1345
 22-    MAT1    8888    30.0+6          .29     7.0-4
 23-    PBAR    5555    8888    4.00            1.333
        ENDDATA
```

Exhibit 19.8 (continued) Static analysis

```
*** USER INFORMATION MESSAGE 5293 FOR DATA BLOCK KLL

LOAD SEQ. NO.        EPSILON         EXTERNAL WORK      EPSILONS LARGER THAN 0.001 ARE FLAGGED WITH ASTERISKS
     1          -5.0053197E-15       2.1567287E-02


                        F O R C E S   I N   B A R   E L E M E N T S        ( C B A R )

ELEMENT     BEND-MOMENT END-A        BEND-MOMENT END-B        - SHEAR -                 AXIAL
  ID.     PLANE 1    PLANE 2       PLANE 1    PLANE 2       PLANE 1    PLANE 2         FORCE      TORQUE
  1001    0.0        0.0           0.0       -1.283450E+02   0.0       2.566900E+01     0.0        0.0
```

Exhibit 19.8 (continued)   Static analysis

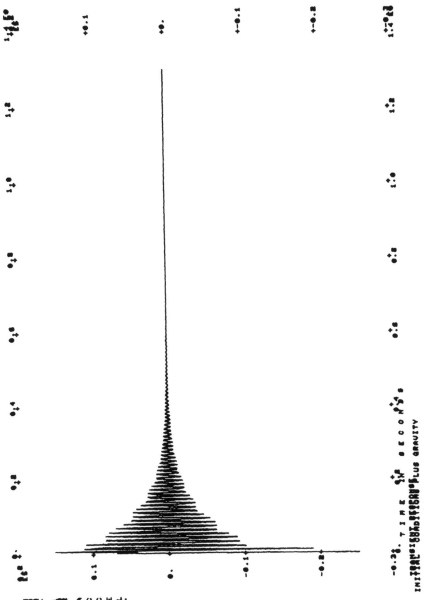

Figure 19.2 Acceleration at Grid 102 as a function of time.

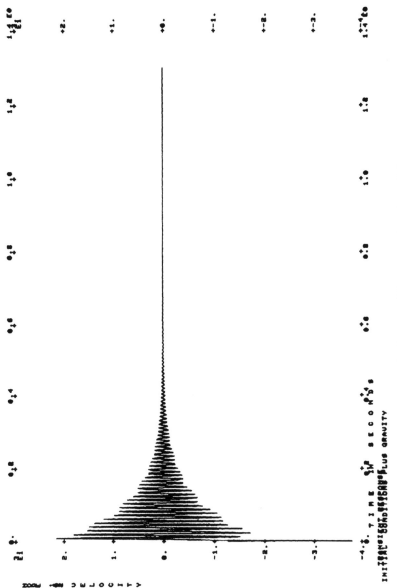

Figure 19.3 Velocity at Grid 102 as a function of time.

20
Problem 20

20.1 Statement of the problem

Determine the response of the structure of Problem 14 (Figure 14.1) to the following loads (expressed in pounds) applied at GRIDs 105 and 108:

$$F_{105}(t) = 2e^{-2t}\cos(290\pi t + \pi/4) \qquad \text{for } t > 0 \qquad (20.1)$$

and,

$$F_{108}(t) = \begin{cases} 0 \text{ if } t < 0.01 \text{ s} \\ \\ 3e^{-t}\cos(290\pi[t-0.01]) \quad \text{for } t > 0.01 \text{ s or } t = 0.01 \text{ s} \end{cases} \qquad (20.2)$$

A 4% fraction of critical viscous damping is assumed for all modes.

20.2 Cards introduced

Case Control Deck   SDAMPING

Bulk Data Deck   DELAY
                                  DLOAD
                                  TABDMP1
                                  TLOAD2
                                  PARAM, HFREQ, xxx

20.3 MSC/NASTRAN formulation

This is a transient response problem, but unlike the previous examples (Problems 18 and 19) we do not use a direct method (SOL 27) to solve the equation of motion. Instead, a modal decomposition method is used, which means that the equation of motion of the system

$$[M]d^2\{x\}/dt^2 + [K]d\{x\}/dt = \{f(t)\} \tag{20.3}$$

is solved by expanding the solution $\{x\}$ as

$$\{x\} = \{\phi_1\}\xi_1 + \{\phi_2\}\xi_2 + \{\phi_3\}\xi_3 + \ldots \tag{20.4}$$

where $\{\phi_i\}$ represents the i-th normal mode and ξ_i represents the i-th generalized coordinate which results from solving an equation of the form

$$d^2\xi_i/dt^2 + 2\omega_i\varsigma_i d\xi_i/dt + \omega^2_i\xi_i = g_i \tag{20.5}$$

where ω_i is the i-th natural frequency, g_i is the i-th generalized force, and ς_i is the fraction of critical viscous damping assumed for the i-th mode.

Transient response using a modal decomposition method is selected with SOL 31 in the Executive Control Deck.

20.4 Input Data Deck

The Executive Control Deck is shown in Exhibit 20.1 and the Case Control Deck is shown in Exhibit 20.2. The Case Control Deck includes a DLOAD card to select the dynamic loads from the Bulk Data Deck, and a TSTEP card which is always required in transient response, whether it is SOL 27 or 31.

The SDAMPING card points to a TABDMP1 card in the Bulk Data Deck. This card specifies the fraction of critical damping to be used for each mode. The METHOD card, used to select an EIGR card for eigenvalue extraction from the Bulk Data Deck, is always required in SOL 31. The reason is that the normal modes and natural frequencies of the system (at least some of them) need to be calculated before expanding the solution using the contribution of each mode.

The plot subdeck states that the displacement at GRID 102, the external load (excitation) at GRID 105, and the external load at GRID 108, are all to be plotted as functions of time.

Examine the Bulk Data Deck (Exhibits 20.3 and 20.4). The DLOAD = 100 statement in the Case Control Deck selects the DLOAD card shown in the Bulk Data Deck. This DLOAD card is used to combine loads in dynamic problems. It can be used in SOL 31, SOL 27, and in frequency response (which will be treated in Problem 21). The DLOAD card refers to the TLOAD2,1,100,101,... and TLOAD2,2,200,201,... cards. In general, the DLOAD card can be used to combine any array of TLOAD1 and TLOAD2 cards.

The TLOAD2 cards are used to define dynamic loads that can be expressed using a particular analytical expression (see Appendix II). It can be noticed that the form of the dynamic loads of this example fits this expression. Note also that the phase angle P in the TLOAD2 card must be in degrees.

The EIGR card selects the modified Householder technique (any other eigenvalue extraction method would also be acceptable) to determine the normal modes and natural frequencies of the system. The first six normal modes are requested.

The TABDMP1 card, selected by SDAMPING = 555 in the Case Control Deck, specifies the modal damping as a function of the frequency. This value is used when solving Eq. 20.5 for the generalized coordinates ξ_i. The keyword CRIT in the third field of the TABDMP1 card indicates that the damping is given as a fraction of the critical damping. This is the most common way of specifying the amount of viscous damping to be considered for a particular mode. The continuation card defines a constant function equal to .04. Thus, when computing ξ_i (Eq. 20.5) it is assumed that for all modes --regardless of their natural frequency-- the fraction of viscous damping ζ_i equals 0.04 (four per cent).

The PARAM, HFREQ, 500.00 card indicates that only modes with natural frequencies below 500.00 Hz will be used in the modal expansion (Eq. 20.4) to determine the response $\{x\}$ of the structure. This is adequate since three normal modes of this structure have natural frequencies below 500.00 Hz, as shown in Exhibit 14.6.

The TSTEP card indicates that 500 steps with a $\Delta t = 0.0003$ s are carried out in the integration process to compute ξ_i (Eq. 20.5).

Finally, note that the PBAR card gives the coordinates of four points on the cross section of the beam where the stresses are to be calculated. This is necessary due to the stress output request made in the Case Control Deck (STRESS = 444).

20.5 Results

Exhibit 20.5 displays the natural frequencies of the structure (they coincide with the values determined in Problems 14 and 15).

Since it was chosen to expand the solution using only those modes whose frequencies were below 500 Hz, only the first three modes are considered in the determination of $\{x\}$ and in the subsequent recovery variables. Note also that the period of the third mode is $1/461.46 = .0021$ s. This is about seven times larger than the time step chosen (3×10^{-4} s), which is appropriate since it indicates that there is consistency between these two selections. In other words, the time step is small enough to capture the influence of the third mode in the response of the beam.

Exhibit 20.6 shows the acceleration at GRID 102 as a function of time. Exhibit 20.7 shows the stresses in BAR element 1004 as a function of time for the four points specified on the PBAR card. (A refers to the left GRID, 104 in this case; B refers to the GRID at the right end of the element, 105 in this case)

The first plot (Figure 20.1) depicts the displacement at GRID 102 as a function of time; the next plots (Figures 20.2 and 20.3) show the external loads applied at GRIDs 105 and 108, respectively.

```
N A S T R A N   E X E C U T I V E   C O N T R O L   D E C K   E C H O

   ID   MODAL,METHOD
   TIME  5
   SOL   31    $  TRANSIENT RESPONSE WITH MODAL  METHOD
   CEND
```

Exhibit 20.1   Executive Control Deck

```
                    C A S E   C O N T R O L   D E C K   E C H O
COUNT
  1      TITLE= TRANSIENT RESPONSE; MODAL METHOD
  2      $$
  3      DLOAD=  100  $ TO SELECT THE LOAD ACTING AT NODE 105  AND  108
  4      $$
  5      TSTEP=200    $ TO SELECT THE TSTEP  CARD
  6      $$
  7      SET  1000=  102
  8      SET 444=  1004
  9      STRESS =444
 10      ACCELERATION = 1000
 11      $$
 12      ECHO=BOTH
 13      $$
 14      SDAMPING= 555 $ TO CHOOSE A DAMPING TABLE
 15      $$
 16      METHOD= 28    $ TO SPECIFY HOW TO COMPUTE THE EIGENVALUES
 17      $$
 18      $$   PLOTS
 19      $$
 20      OUTPUT(XYPLOT)
 21      XTAXIS=YES
 22      XBAXIS=YES
 23      XTGRID  LINES=  YES
 24      XBGRID  LINES = YES
 25      YTGRID LINES = YES
 26      YBGRID LINES = YES
 27      $$
 28      XTITLE=  T I M E   IN   S E C O N D S
 29      YTITLE=  NODE  102   DISPLACEMENT
 30      XYPLOT  DISP   /102(T2)
 31      $$
 32      YTITLE=  EXTERNAL LOAD   AT 105
 33      XYPLOT  OLOAD  /105(T2)
 34      $$
 35      YTITLE= EXTERNAL LOAD   AT 108
 36      XYPLOT  OLOAD /108(T2)
 37      $$
 38      BEGIN BULK
```

Exhibit 20.2 Case Control Deck

```
            I N P U T   B U L K   D A T A   D E C K   E C H O
      .   1  ..  2  ..  3  ..  4  ..  5  ..  6  ..  7  ..  8  ..  9  ..  10  .
$$
$$$$  DYNAMIC LOAD
$$
DLOAD    100     1.00    1.00       1    1.00
TLOAD2     1      100     101       0    .000   10.+9    145.    45.    +BBB
+BBB      -2.      .0
DAREA    100      105       2    2.00
DELAY    101      105       2    .000
$$
TLOAD2     2      200     201       0    .000   10.+9    145.    .00    +TTT
+TTT      -1.      .0
DAREA    200      108       2    3.00
DELAY    201      108       2     .01
$$$$
$$$$  EIGENVALUE EXTRACTION
$$$$
EIGR      28     MHOU                             6
$$
$$  DAMPING TABLE
$$
TABDMP1  555     CRIT
+KKKK     .00     .04   5555.     .04    ENDT                          +KKKK
$$$$
```

Exhibit 20.3   Input Bulk Data Deck

```
.   1  ..  2  ..  3  ..  4  ..  5  ..  6  ..  7  ..  8  ..  9  ..  10  .
    INPUT  BULK  DATA  DECK  ECHO

$$$$  TO EXPAND THE SOLUTION USING ONLY

$$$$  MODES WHOSE FREQ. IS BELOW  500 HZ.

$$$$

PARAM,HFREQ,500.00

$$$$

$$$$$$$$$$$$$$$$$$$$$$$$$$$$$$$$$$$

$$  NUMERICAL INTEGRATION PARAMETERS

TSTEP   200    500    3.-4    2

$$$$$$$$$$$$$$$$$$$$$$$$$$$$$$$$$$$

GRID    101    00.00    0.00    1345
GRID    102    5.       0.00    1345
GRID    103    10.00    0.00    1345
GRID    104    15.      0.00    1345
GRID    105    20.00    0.00    1345
GRID    106    25.00    0.00    1345
GRID    107    30.00    0.00    1345
GRID    108    35.00    0.00    1345
GRID    109    40.      0.00    1345
```

Exhibit 20.3 (continued)   Input Bulk Data Deck

```
            I N P U T   B U L K   D A T A   D E C K   E C H O

      . 1 ..  2  ..  3  ..  4  ..  5  ..  6  ..  7  ..  8  ..  9  ..  10 .
CBAR    1001 5555   101    102   1.000  0.000  1.000
CBAR    1002 5555   102    103   1.000  0.000  1.000
CBAR    1003 5555   103    104   1.000  0.000  1.000
CBAR    1004 5555   104    105   1.000  0.000  1.000
CBAR    1005 5555   105    106   1.000  0.000  1.000
CBAR    1006 5555   106    107   1.000  0.000  1.000
CBAR    1007 5555   107    108   1.000  0.000  1.000
CBAR    1008 5555   108    109   1.000  0.000  1.000
$$
PBAR    5555 8888  4.00          1.333                            +J
+J      2.0  2.    2.    -2.     -2.    -2.    -2.    2.
$$$ WE HAVE INCLUDED THE COORDINATES OF
$$$ THE POINTS WHERE WE WANT THE STRESSES
$$$
MAT1    8888 30.0+6       .29    7.0-4
$$ CONCENTRATED MASS    AT POINT 102
CMASS2  123456 .02    102    2
$$ SPRINGS AT NODE 101    AND  109
CELAS2  2233   10.+4  101    2
CELAS2  2244   10.+4  109    2
ENDDATA
```

Exhibit 20.3 (continued) Input Bulk Data Deck

S O R T E D   B U L K   D A T A   E C H O

CARD COUNT	. 1	.. 2	.. 3	.. 4	.. 5	.. 6	.. 7	.. 8	.. 9	.. 10 .
1-	CBAR	1001	5555	101	102	1.000	0.000	1.000		
2-	CBAR	1002	5555	102	103	1.000	0.000	1.000		
3-	CBAR	1003	5555	103	104	1.000	0.000	1.000		
4-	CBAR	1004	5555	104	105	1.000	0.000	1.000		
5-	CBAR	1005	5555	105	106	1.000	0.000	1.000		
6-	CBAR	1006	5555	106	107	1.000	0.000	1.000		
7-	CBAR	1007	5555	107	108	1.000	0.000	1.000		
8-	CBAR	1008	5555	108	109	1.000	0.000	1.000		
9-	CELAS2	2233	10.+4	101	2					
10-	CELAS2	2244	10.+4	109	2					
11-	CMASS2	123456	.02	102	2					
12-	DAREA	100	105	2	2.00					
13-	DAREA	200	108	2	3.00					
14-	DELAY	101	105	2	.000					
15-	DELAY	201	108	2	.01					
16-	DLOAD	100	1.00	1.00	1	1.00	2			
17-	EIGR	28	MHOU				6			

Exhibit 20.4   Bulk Data Deck sorted by alphabetical order

```
18-  GRID     101              00.00   0.00    0.00            1345
19-  GRID     102              5.      0.00    0.00            1345
20-  GRID     103              10.00   0.00    0.00            1345
21-  GRID     104              15.     0.00    0.00            1345
22-  GRID     105              20.00   0.00    0.00            1345
23-  GRID     106              25.00   0.00    0.00            1345
24-  GRID     107              30.00   0.00    0.00            1345
25-  GRID     108              35.00   0.00    0.00            1345
26-  GRID     109              40.     0.00    0.00            1345
27-  MAT1     8888     30.0+6          .29     7.0-4
28-  PARAM    HFREQ    500.00
29-  PBAR     5555     8888     4.00            1.333                           +J
30-  +J       2.0      2.       2.      -2.     -2.             -2.     2.
31-  TABDMP1  555      CRIT                                                     +KKKK
32-  +KKKK    .00      .04      .04     -2.     2.
33-  TLOAD2   1        100      5555.   0       ENDT    10.+9   145.    45.     +BBB
34-  +BBB     -2.      .0
35-  TLOAD2   2        200      201     0       .000    10.+9   145.    .00     +TTT
36-  +TTT     -1.      .0
37-  TSTEP    200      500      3.-4    2
     ENDDATA
```

Exhibit 20.4 (continued)   Bulk Data Deck sorted by alphabetical order

R E A L   E I G E N V A L U E S

MODE NO.	EXTRACTION ORDER	EIGENVALUE	RADIANS	CYCLES	GENERALIZED MASS	GENERALIZED STIFFNESS
1	1	4.011898E+05	6.333954E+02	1.008080E+02	1.000000E+00	4.011898E+05
2	2	2.870254E+06	1.694182E+03	2.696375E+02	1.000000E+00	2.870254E+06
3	3	8.406914E+06	2.899468E+03	4.614646E+02	1.000000E+00	8.406914E+06
4	4	2.469761E+07	4.969669E+03	7.909474E+02	1.000000E+00	2.469761E+07
5	5	6.807676E+07	8.250864E+03	1.313166E+03	1.000000E+00	6.807676E+07
6	6	1.525866E+08	1.235259E+04	1.965976E+03	1.000000E+00	1.525866E+08
7	7	3.208477E+08	1.791222E+04	2.850819E+03	0.0	0.0
8	8	6.047239E+08	2.459113E+04	3.913801E+03	0.0	0.0
9	9	9.357556E+08	3.059012E+04	4.868569E+03	0.0	0.0

Exhibit 20.5   Eigenvalues

POINT-ID =   102

			ACCELERATION	VECTOR			
TIME	TYPE	T1	T2	T3	R1	R2	R3
0.0	G	0.0	-1.171410E+00	0.0	0.0	0.0	3.264004E+00
6.000000E-04	G	0.0	7.413884E+00	0.0	0.0	0.0	-7.463491E-01
1.200000E-03	G	0.0	4.019993E+00	0.0	0.0	0.0	-2.298804E+00
1.800000E-03	G	0.0	-1.543437E+01	0.0	0.0	0.0	-8.199415E-02
2.400000E-03	G	0.0	-2.051726E+01	0.0	0.0	0.0	-4.643063E-01
3.000000E-03	G	0.0	-1.096769E+01	0.0	0.0	0.0	-2.110198E+00
3.600000E-03	G	0.0	-5.228078E+00	0.0	0.0	0.0	8.494826E-02
4.200000E-03	G	0.0	5.548012E+00	0.0	0.0	0.0	2.936401E+00
4.800000E-03	G	0.0	2.855561E+01	0.0	0.0	0.0	2.363006E+00
5.400000E-03	G	0.0	4.034599E+01	0.0	0.0	0.0	1.762803E+00
6.000000E-03	G	0.0	2.864647E+01	0.0	0.0	0.0	3.081549E+00

SOME LINES HAVE BEEN DELETED

1.464000E-01	G	0.0	-1.071959E-01	0.0	0.0	0.0	-1.004761E+00
1.470000E-01	G	0.0	-4.988612E+00	0.0	0.0	0.0	-5.597519E-01
1.476000E-01	G	0.0	2.110998E+00	0.0	0.0	0.0	3.823585E-02
1.482000E-01	G	0.0	8.489583E+00	0.0	0.0	0.0	6.127145E-01
1.488000E-01	G	0.0	1.228305E+01	0.0	0.0	0.0	9.959667E-01
1.494000E-01	G	0.0	1.239858E+01	0.0	0.0	0.0	1.078048E+00
1.500001E-01	G	0.0	8.832509E+00	0.0	0.0	0.0	8.385417E-01

Exhibit 20.6   Acceleration vector as a function of time

ELEMENT-ID =   1004

TIME	SA1 SB1	S T R E S S E S	SA2 SB2	SA3 SB3 (I N B A R	SA4 SB4 E L E M E N T S)	AXIAL- STRESS	(C B A R) SA-MAX SB-MAX	SA-MIN SB-MIN	M.S.-T M.S.-C
0.0	0.0 0.0	0.0 0.0	0.0 0.0	0.0 0.0	0.0 0.0	0.0 0.0	0.0 0.0	0.0 0.0	
6.000000E-04	-2.374791E+00 -2.805560E+00		2.374791E+00 2.805560E+00	2.374791E+00 2.805560E+00	-2.374791E+00 -2.805560E+00	0.0	2.374791E+00 2.805560E+00	-2.374791E+00 -2.805560E+00	
1.200000E-03	-2.817712E+00 -3.416331E+00		2.817712E+00 3.416331E+00	2.817712E+00 3.416331E+00	-2.817712E+00 -3.416331E+00	0.0	2.817712E+00 3.416331E+00	-2.817712E+00 -3.416331E+00	
1.800000E-03	1.083846E+00 1.331812E+00		-1.083846E+00 -1.331812E+00	-1.083846E+00 -1.331812E+00	1.083846E+00 1.331812E+00	0.0	1.083846E+00 1.331812E+00	-1.083846E+00 -1.331812E+00	

S O M E   L I N E S   H A V E   B E E N   D E L E T E D

TIME	SA1 SB1	SA2 SB2	SA3 SB3	SA4 SB4	AXIAL- STRESS	SA-MAX SB-MAX	SA-MIN SB-MIN	M.S.-T M.S.-C
1.482000E-01	1.043722E+01 6.390220E+00	-1.043722E+01 -6.390220E+00	-1.043722E+01 -6.390220E+00	1.043722E+01 6.390220E+00	0.0	1.043722E+01 6.390220E+00	-1.043722E+01 -6.390220E+00	
1.488000E-01	1.074666E+01 7.602160E+00	-1.074666E+01 -7.602160E+00	-1.074666E+01 -7.602160E+00	1.074666E+01 7.602160E+00	0.0	1.074666E+01 7.602160E+00	-1.074666E+01 -7.602160E+00	
1.494000E-01	7.801780E+00 6.468388E+00	-7.801780E+00 -6.468388E+00	-7.801780E+00 -6.468388E+00	7.801780E+00 6.468388E+00	0.0	7.801780E+00 6.468388E+00	-7.801780E+00 -6.468388E+00	
1.500001E-01	2.497725E+00 3.357460E+00	-2.497725E+00 -3.357460E+00	-2.497725E+00 -3.357460E+00	2.497725E+00 3.357460E+00	0.0	2.497725E+00 3.357460E+00	-2.497725E+00 -3.357460E+00	

Exhibit 20.7   Stresses in BAR elements as a function of time

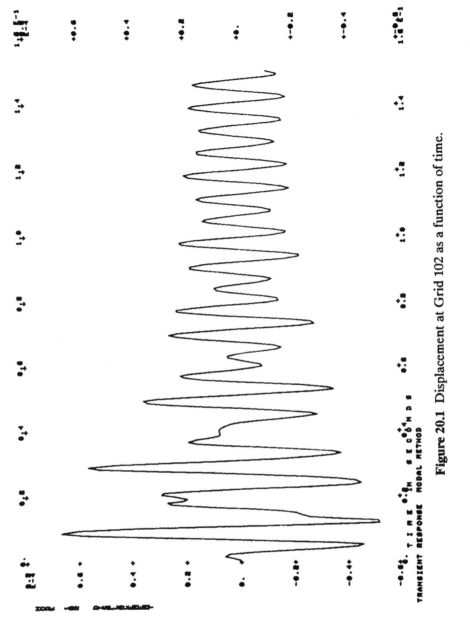

Figure 20.1 Displacement at Grid 102 as a function of time.

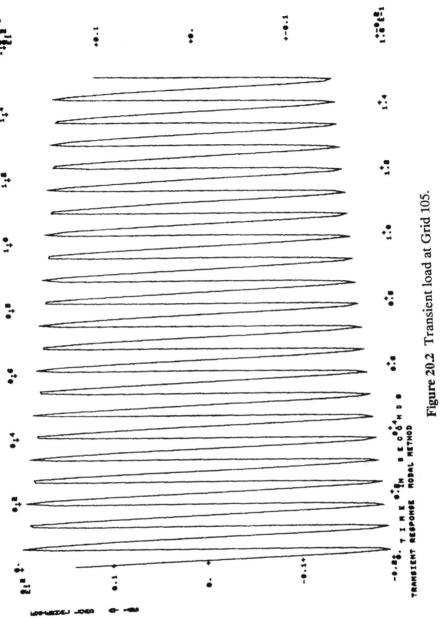

Figure 20.2 Transient load at Grid 105.

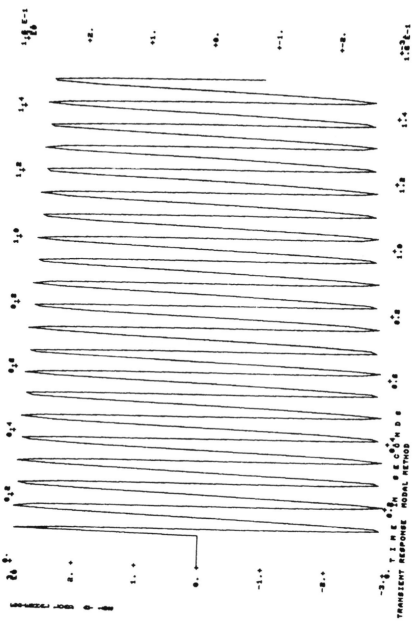

Figure 20.3 Transient load at Grid 108.

21

Problem 21

21.1 Statement of the problem

Consider the structure of Problem 14 (Figures 14.1 and 14.2) and assume that two dashpots (viscous dampers) with a constant $c = 1.5$ lb-s/in have been added at both ends (as in Problems 18 and 19). The structure is excited by a harmonic force F (in pounds) given by

$$F(f) = P(f)e^{i(2\pi f)t} = 45e^{i(2\pi f)t} \tag{21.1}$$

applied at GRID 105 in the y-direction. In this equation, i represents the square root of -1; f is the frequency of the excitation in Hz; t represents the time in seconds, and P(f) is the frequency dependent amplitude of the harmonic excitation. Compute the steady-state response of the beam for frequencies of excitation in the range $f = 80$ Hz to $f = 160$ Hz.

21.2 Cards introduced

Case Control Deck   FREQUENCY

Bulk Data Deck   FREQ1
                                     RLOAD1

21.3 MSC/NASTRAN formulation

The structure is modeled as in Problems 18 and 19. This is a direct frequency response analysis, which corresponds to SOL 26. This analysis consists of applying to the structure a harmonic force (or a combination of forces) of the form

$$F(f) = P(f)e^{i(2\pi f)t} \tag{21.2}$$

and determining the steady-state response for each value of the forcing frequency (f). The amplitude of the harmonic force (P(f)) is defined with an RLOAD1 card.

21.4 Input Data Deck

The Executive Control Deck and the Case Control Deck are shown in Exhibits 21.1 and 21.2, respectively.

A few comments regarding the Case Control Deck are in order. The DLOAD card refers to an RLOAD1 card in the Bulk Data Deck that defines the amplitude of the harmonic force. This DLOAD card could also have pointed to another DLOAD card in the Bulk Data Deck, which in turn could have combined several RLOAD1 and RLOAD2 cards if several harmonic loads were to be considered at the same time.

The DISP card in frequency response analysis is used to request the steady-state value of the displacement to be printed. In this example, it requests the displacement for GRID 102. The displacement is computed for several excitation frequencies. The choice SORT2 made with the DISP card requests that the output be sorted by frequency (as opposed to by GRID point which is the default). In addition, notice that the recovery variables are complex numbers. The selection PHASE on the DISP card prints the output in terms of amplitude and phase instead of real and imaginary parts, which is the default option.

The FREQUENCY card points to a FREQ1 card in the Bulk Data Deck.

The first plot request, XYPLOT DISP /102(T2RM), plots the magnitude of the displacement at GRID 102 (y-direction) as a function of the frequency of excitation. The string RM after T2 can mean two things, "Real" (R) or "Magnitude" (M). Since the choice on the DISP card is to print the results in terms of magnitude and phase, RM in this context means "Magnitude".

The second plot request, XYPLOT DISP /102(T2IP), plots the phase angle for the displacement response at GRID 102 as a function of frequency. The IP after T2, which in general can refer to "Imaginary" or "Phase", corresponds to "Phase" in this context, again consistent with the choice of PHASE on the DISP card.

Examine the RLOAD1 card in the Bulk Data Deck (Exhibits 21.3 and 21.4) which is employed to define *only* the amplitude of the harmonic excitation (P(f)). The harmonic nature of this function, given by the term $e^{i(2\pi f)t}$ in Eq. (21.1), is implicitly defined with the RLOAD1 card. This card has an ID=100 consistent with DLOAD=100 in the Case Control Deck. It also points to a DAREA (ID=101) card and a TABLED1 (ID=55). The DAREA card specifies that the harmonic excitation is applied at GRID 105 in the y-direction (2) with a scale factor of 45. The TABLED1 card defines a constant function equal to 1.0 for all frequencies. Thus, C(f) = 1.0 for all values of f. No entries in the fourth, fifth, and seventh field of the RLOAD1 card are specified, which means that τ, θ, and D(f) are zero. In summary, the RLOAD1, DAREA and TABLED1 cards of this example define a harmonic function of constant amplitude (45) applied at GRID 105, consistent with Eq. 21.1. (see Appendix II). That is

$$P(f) = \{45. [1.0 + i\, 0.0]\, e^{i\{0-0\}}\} = 45.0 \qquad (21.3)$$

The FREQ1 card, selected in the Case Control Deck by FREQUENCY = 10, specifies that the load defined by Eq. 21.1 is to be applied for values of f starting at f = 80.0 Hz and increasing by increments of 1.0 Hz eighty times. The rest of the Bulk Data Deck should be familiar at this point.

21.5 Results

Exhibit 21.5 shows the output generated by DISP(SORT2, PHASE) = 1000. It lists the magnitude of the steady-state displacement and its phase angle as a function of the frequency of excitation, at GRID 102. Recall that the fundamental frequency of this structure is 100.8 Hz (Problems 14 and 15). Therefore, a peak in the response should appear when the frequency of the excitation approaches this value.

At f = 101 Hz a local maximum (peak) in the value of the displacement is observed. Moreover, the phase angle changes rapidly from about 360° (before the peak) to roughly 180° (after the peak), consistent with what one expects from the theory. The first plot (Figure 21.1) shows the magnitude of the displacement as a function of the frequency and clearly indicates a maximum around 100.00 Hz. The second plot (Figure 21.2) shows the variation of the phase angle as a function of the frequency of excitation. This observed behavior is consistent with what one would intuitively expect for this structure.

```
N A S T R A N   E X E C U T I V E   C O N T R O L   D E C K   E C H O
    ID   A,B
    TIME  5
    SOL   26  $ FREQ. RESPONSE  ( DIRECT METHOD )
    CEND
```

Exhibit 21.1   Executive Control Deck

```
                    C A S E   C O N T R O L   D E C K   E C H O

CARD
COUNT

  1    TITLE= FREQ.  RESPONSE

  2    $$$

  3    DLOAD=  100  $ TO SELECT THE LOAD ACTING AT NODE 105

  4    $$$

  5    $

  6    $$$

  7    SET  1000=  102

  8    DISP(SORT2,PHASE)  =  1000   $   IMPORTANT:   SORT2, MEANS THAT THE RESULTS WILL

  9    $$                                           SORTED BY FREQ. FOR EACH GRID POINT

 10    $$

 11    $$                                           PHASE, MEANS THAT THE RESULTS WILL

 12    $$                                           PRINTED IN TERMS OF AMPLITUDE AND P

 13    $$

 14    $$

 15    ECHO=BOTH

 16    $$

 17    FREQUENCY =10  $$  TO SELECT A FREQ.   CARD
```

Exhibit 21.2   Case Control Deck

```
                   C A S E    C O N T R O L    D E C K    E C H O

CARD
COUNT
  18    $$
  19    $$   PLOTS   FOR  FREQUENCY   RESPONSE
  20    $$
  21    OUTPUT(XYPLOT)
  22    XGRID=YES
  23    YGRID=YES
  24    XTITLE=  F R E Q U E N C Y    I N    H Z .
  25    YTITLE=  M A G N I T U D E
  26    XYPLOT  DISP  /102(T2RM)
  27    $$
  28    YTITLE=  P H A S E
  29    XYPLOT  DISP  /102(T2IP)
  30    BEGIN BULK
```

Exhibit 21.2 (continued)   Case Control Deck

```
            I N P U T   B U L K   D A T A   D E C K   E C H O
.   1 .. 2 .. 3 .. 4 .. 5 .. 6 .. 7 .. 8 .. 9 .. 10 .
$$
$$ FREQ. RANGE
$$
FREQ1    10     80.    1.00   80
$$$$ DYNAMIC LOAD
$$
RLOAD1   100    101
DAREA    101    105    2      45.
TABLED1  55                   55
+V       .00    1.00   10.+9  1.     ENDT                              +V
$$
$$ DASHPOTS (DAMPERS)
CDAMP2   4455   1.5    101    2
CDAMP2   3322   1.5    109    2
$
GRID     101    00.00         0.00   0.00                      1345
GRID     102    5.            0.00   0.00                      1345
GRID     103    10.00         0.00   0.00                      1345
GRID     104    15.           0.00   0.00                      1345
GRID     105    20.00         0.00   0.00                      1345
GRID     106    25.00         0.00   0.00                      1345
GRID     107    30.00         0.00   0.00                      1345
GRID     108    35.00         0.00   0.00                      1345
GRID     109    40.           0.00   0.00                      1345
```

Exhibit 21.3   Input Bulk Data Deck

```
           I N P U T   B U L K   D A T A   D E C K   E C H O
  . 1 ..  2 ..   3 ..   4 ..  5 ..  6 ..    7 ..   8 ..   9 ..  10 .
  CBAR    1001 5555    101   102   1.000   0.000  1.000
  CBAR    1002 5555    102   103   1.000   0.000  1.000
  CBAR    1003 5555    103   104   1.000   0.000  1.000
  CBAR    1004 5555    104   105   1.000   0.000  1.000
  CBAR    1005 5555    105   106   1.000   0.000  1.000
  CBAR    1006 5555    106   107   1.000   0.000  1.000
  CBAR    1007 5555    107   108   1.000   0.000  1.000
  CBAR    1008 5555    108   109   1.000   0.000  1.000
  PBAR    5555 8888   4.00         1.333
  MAT1    8888 30.0+6  .29   7.0-4
  $$ CONCENTRATED MASS    AT POINT 102
  CMASS2 123456 .02    102   2
  $$ SPRINGS AT NODE  101    AND 109
  CELAS2 2233  10.+4   101   2
  CELAS2 2244  10.+4   109   2
  ENDDATA
```

Exhibit 21.3 (continued) Input Bulk Data Deck

S O R T E D   B U L K   D A T A   E C H O

CARD	. 1	.. 2	.. 3	.. 4	.. 5	.. 6	.. 7	.. 8	.. 9	.. 10 .
1-	CBAR	1001	5555	101	102	1.000	0.000	1.000		
2-	CBAR	1002	5555	102	103	1.000	0.000	1.000		
3-	CBAR	1003	5555	103	104	1.000	0.000	1.000		
4-	CBAR	1004	5555	104	105	1.000	0.000	1.000		
5-	CBAR	1005	5555	105	106	1.000	0.000	1.000		
6-	CBAR	1006	5555	106	107	1.000	0.000	1.000		
7-	CBAR	1007	5555	107	108	1.000	0.000	1.000		
8-	CBAR	1008	5555	108	109	1.000	0.000	1.000		
9-	CDAMP2	3322	1.5	109	2					
10-	CDAMP2	4455	1.5	101	2					
11-	CELAS2	2233	10.+4	101	2					
12-	CELAS2	2244	10.+4	109	2					
13-	GMASS2	123456	.02	102	2					
14-	DAREA	101	105	2	45.					
15-	FREQ1	10	80.	1.00	80					
16-	GRID	101		00.00		0.00	0.00		1345		
17-	GRID	102		5.		0.00	0.00		1345		
18-	GRID	103		10.00		0.00	0.00		1345		
19-	GRID	104		15.		0.00	0.00		1345		
20-	GRID	105		20.00		0.00	0.00		1345		
21-	GRID	106		25.00		0.00	0.00		1345		
22-	GRID	107		30.00		0.00	0.00		1345		
23-	GRID	108		35.00		0.00	0.00		1345		
24-	GRID	109		40.		0.00	0.00		1345		

Exhibit 21.4   Bulk Data Deck sorted by alphabetical order

```
                        S O R T E D   B U L K   D A T A   E C H O

CARD
COUNT    .  . 1 . .  2  . . 3  . . 4 . . 5  . . 6 . . 7 . . 8 . . 9 . . 10 .
25-        MAT1   8888   30.0+6    .29     7.0-4
26-        PBAR   5555   8888     4.00     1.333
27-        RLOAD1 100    101               55
28-        TABLED1 55                                                  +V
29-        +V      .00   1.00    10.+9   1.        ENDT
30-        TSTEP  200    1300     3.-4    2
           ENDDATA
```

Exhibit 21.4 (continued)   Bulk Data Deck sorted by alphabetical order

POINT-ID =   102

C O M P L E X   D I S P L A C E M E N T   V E C T O R
(MAGNITUDE/PHASE)

FREQUENCY	TYPE	T1	T2	T3	R1	R2	R3
8.000000E+01	G	0.0	2.221688E-03	0.0	0.0	0.0	2.751466E-04
		0.0	359.6887	0.0	0.0	0.0	359.8543
8.100000E+01	G	0.0	2.324549E-03	0.0	0.0	0.0	2.871742E-04
		0.0	359.6720	0.0	0.0	0.0	359.8409
8.200000E+01	G	0.0	2.438514E-03	0.0	0.0	0.0	3.004978E-04
		0.0	359.6537	0.0	0.0	0.0	359.8258

Exhibit 21.5   Complex displacement vector

8.300000E+01	G	0.0	0.0	2.565450E-03	0.0	0.0	0.0	3.153354E-04	0.0
			0.0	359.6335	0.0		0.0	359.8089	0.0
8.400000E+01	G	0.0	0.0	2.707671E-03	0.0	0.0	0.0	3.319568E-04	0.0
			0.0	359.6110	0.0		0.0	359.7898	0.0
8.500001E+01	G	0.0	0.0	2.868074E-03	0.0	0.0	0.0	3.507001E-04	0.0
			0.0	359.5857	0.0		0.0	359.7679	0.0
8.600000E+01	G	0.0	0.0	3.050338E-03	0.0	0.0	0.0	3.719444E-04	0.0
			0.0	359.5572	0.0		0.0	359.7429	0.0
8.700000E+01	G	0.0	0.0	3.259212E-03	0.0	0.0	0.0	3.963941E-04	0.0
			0.0	359.5247	0.0		0.0	359.7139	0.0
8.800001E+01	G	0.0	0.0	3.500925E-03	0.0	0.0	0.0	4.246258E-04	0.0
			0.0	359.4873	0.0		0.0	359.6801	0.0
8.900000E+01	G	0.0	0.0	3.783811E-03	0.0	0.0	0.0	4.576618E-04	0.0
			0.0	359.4437	0.0		0.0	359.6401	0.0
9.000000E+01	G	0.0	0.0	4.119298E-03	0.0	0.0	0.0	4.968357E-04	0.0
			0.0	359.3923	0.0		0.0	359.5924	0.0
9.100000E+01	G	0.0	0.0	4.523467E-03	0.0	0.0	0.0	5.440236E-04	0.0
			0.0	359.3306	0.0		0.0	359.5345	0.0
9.200001E+01	G	0.0	0.0	5.019694E-03	0.0	0.0	0.0	6.019532E-04	0.0
			0.0	359.2552	0.0		0.0	359.4630	0.0
9.300000E+01	G	0.0	0.0	5.643328E-03	0.0	0.0	0.0	6.747487E-04	0.0
			0.0	359.1609	0.0		0.0	359.3724	0.0
9.400000E+01	G	0.0	0.0	6.450488E-03	0.0	0.0	0.0	7.689580E-04	0.0
			0.0	359.0391	0.0		0.0	359.2546	0.0
9.500001E+01	G	0.0	0.0	7.535886E-03	0.0	0.0	0.0	8.956316E-04	0.0
			0.0	358.8759	0.0		0.0	359.0954	0.0

Exhibit 21.5 (continued)   Complex displacement vector

9.600000E+01	G	0.0 / 0.0	9.072929E-03 / 358.6453	0.0 / 0.0	0.0 / 0.0	1.075002E-03 / 358.8690
9.700000E+01	G	0.0 / 0.0	1.141694E-02 / 358.2945	0.0 / 0.0	0.0 / 0.0	1.348528E-03 / 358.5223
9.800000E+01	G	0.0 / 0.0	1.542786E-02 / 357.6951	0.0 / 0.0	0.0 / 0.0	1.816542E-03 / 357.9272
9.900001E+01	G	0.0 / 0.0	2.385716E-02 / 356.4364	0.0 / 0.0	0.0 / 0.0	2.800072E-03 / 356.6727
1.000000E+02	G	0.0 / 0.0	5.279265E-02 / 352.1024	0.0 / 0.0	0.0 / 0.0	6.176118E-03 / 352.3431
1.010000E+02	G	0.0 / 0.0	1.940436E-01 / 210.2818	0.0 / 0.0	0.0 / 0.0	2.262632E-02 / 210.5269
1.020000E+02	G	0.0 / 0.0	3.581582E-02 / 185.3292	0.0 / 0.0	0.0 / 0.0	4.162387E-03 / 185.5788
1.030000E+02	G	0.0 / 0.0	1.947481E-02 / 182.8872	0.0 / 0.0	0.0 / 0.0	2.255663E-03 / 183.1413
1.040000E+02	G	0.0 / 0.0	1.334362E-02 / 181.9714	0.0 / 0.0	0.0 / 0.0	1.540239E-03 / 182.2302
1.050000E+02	G	0.0 / 0.0	1.013439E-02 / 181.4914	0.0 / 0.0	0.0 / 0.0	1.165751E-03 / 181.7550
1.060000E+02	G	0.0 / 0.0	8.160941E-03 / 181.1956	0.0 / 0.0	0.0 / 0.0	9.354521E-04 / 181.4641
1.070000E+02	G	0.0 / 0.0	6.825011E-03 / 180.9948	0.0 / 0.0	0.0 / 0.0	7.795384E-04 / 181.2681
1.080000E+02	G	0.0 / 0.0	5.860798E-03 / 180.8493	0.0 / 0.0	0.0 / 0.0	6.669956E-04 / 181.1277

Exhibit 21.5 (continued)   Complex displacement vector

1.090000E+02	G	0.0	5.132226E-03	0.0	0.0	5.819469E-04	0.0
		0.0	180.7389	0.0	0.0	181.0224	0.0
1.100000E+02	G	0.0	4.562415E-03	0.0	0.0	5.154214E-04	0.0
		0.0	180.6521	0.0	0.0	180.9408	0.0
1.110000E+02	G	0.0	4.104633E-03	0.0	0.0	4.619668E-04	0.0
		0.0	180.5820	0.0	0.0	180.8759	0.0
1.120000E+02	G	0.0	3.728857E-03	0.0	0.0	4.180800E-04	0.0
		0.0	180.5241	0.0	0.0	180.8234	0.0
1.130000E+02	G	0.0	3.414912E-03	0.0	0.0	3.814068E-04	0.0
		0.0	180.4753	0.0	0.0	180.7801	0.0
1.140000E+02	G	0.0	3.148739E-03	0.0	0.0	3.503069E-04	0.0
		0.0	180.4336	0.0	0.0	180.7440	0.0
1.150000E+02	G	0.0	2.920233E-03	0.0	0.0	3.236015E-04	0.0
		0.0	180.3976	0.0	0.0	180.7136	0.0
1.160000E+02	G	0.0	2.721960E-03	0.0	0.0	3.004227E-04	0.0
		0.0	180.3659	0.0	0.0	180.6878	0.0
1.170000E+02	G	0.0	2.548320E-03	0.0	0.0	2.801176E-04	0.0
		0.0	180.3379	0.0	0.0	180.6656	0.0
1.180000E+02	G	0.0	2.395010E-03	0.0	0.0	2.621838E-04	0.0
		0.0	180.3130	0.0	0.0	180.6467	0.0
1.190000E+02	G	0.0	2.258678E-03	0.0	0.0	2.462303E-04	0.0
		0.0	180.2904	0.0	0.0	180.6302	0.0
1.200000E+02	G	0.0	2.136671E-03	0.0	0.0	2.319475E-04	0.0
		0.0	180.2701	0.0	0.0	180.6161	0.0
1.210000E+02	G	0.0	2.026861E-03	0.0	0.0	2.190871E-04	0.0
		0.0	180.2514	0.0	0.0	180.6038	0.0

Exhibit 21.5 (continued) Complex displacement vector

1.220000E+02	G	0.0	1.927519E-03	0.0	0.0	2.074475E-04
		0.0	180.2343	0.0	0.0	180.5932
1.230000E+02	G	0.0	1.837230E-03	0.0	0.0	1.968633E-04
		0.0	180.2185	0.0	0.0	180.5840
1.240000E+02	G	0.0	1.754824E-03	0.0	0.0	1.871982E-04
		0.0	180.2038	0.0	0.0	180.5760
1.250000E+02	G	0.0	1.679323E-03	0.0	0.0	1.783380E-04
		0.0	180.1902	0.0	0.0	180.5692
1.260000E+02	G	0.0	1.609903E-03	0.0	0.0	1.701866E-04
		0.0	180.1774	0.0	0.0	180.5633
1.270000E+02	G	0.0	1.545869E-03	0.0	0.0	1.626627E-04
		0.0	180.1653	0.0	0.0	180.5584
1.280000E+02	G	0.0	1.486626E-03	0.0	0.0	1.556972E-04
		0.0	180.1539	0.0	0.0	180.5542
1.290000E+02	G	0.0	1.431664E-03	0.0	0.0	1.492303E-04
		0.0	180.1431	0.0	0.0	180.5508
1.300000E+02	G	0.0	1.380544E-03	0.0	0.0	1.432108E-04
		0.0	180.1328	0.0	0.0	180.5481
1.310000E+02	G	0.0	1.332883E-03	0.0	0.0	1.375941E-04
		0.0	180.1230	0.0	0.0	180.5460
1.320000E+02	G	0.0	1.288349E-03	0.0	0.0	1.323414E-04
		0.0	180.1136	0.0	0.0	180.5444
1.330000E+02	G	0.0	1.246651E-03	0.0	0.0	1.274186E-04
		0.0	180.1046	0.0	0.0	180.5434

Exhibit 21.5 (continued)   Complex displacement vector

1.340000E+02	G	0.0 0.0	1.207532E-03 180.0959	0.0 0.0	0.0 0.0	0.0 0.0	1.227959E-04 180.5429
1.350000E+02	G	0.0 0.0	1.170766E-03 180.0875	0.0 0.0	0.0 0.0	0.0 0.0	1.184469E-04 180.5429
1.360000E+02	G	0.0 0.0	1.136152E-03 180.0794	0.0 0.0	0.0 0.0	0.0 0.0	1.143481E-04 180.5432
1.370000E+02	G	0.0 0.0	1.103513E-03 180.0715	0.0 0.0	0.0 0.0	0.0 0.0	1.104787E-04 180.5441
1.380000E+02	G	0.0 0.0	1.072689E-03 180.0638	0.0 0.0	0.0 0.0	0.0 0.0	1.068202E-04 180.5452
1.390000E+02	G	0.0 0.0	1.043538E-03 180.0563	0.0 0.0	0.0 0.0	0.0 0.0	1.033558E-04 180.5468
1.400000E+02	G	0.0 0.0	1.015931E-03 180.0490	0.0 0.0	0.0 0.0	0.0 0.0	1.000707E-04 180.5487
1.410000E+02	G	0.0 0.0	9.897550E-04 180.0418	0.0 0.0	0.0 0.0	0.0 0.0	9.695146E-05 180.5510
1.420000E+02	G	0.0 0.0	9.649046E-04 180.0348	0.0 0.0	0.0 0.0	0.0 0.0	9.398588E-05 180.5536
1.430000E+02	G	0.0 0.0	9.412859E-04 180.0278	0.0 0.0	0.0 0.0	0.0 0.0	9.116295E-05 180.5565
1.440000E+02	G	0.0 0.0	9.188139E-04 180.0210	0.0 0.0	0.0 0.0	0.0 0.0	8.847273E-05 180.5598
1.450000E+02	G	0.0 0.0	8.974110E-04 180.0143	0.0 0.0	0.0 0.0	0.0 0.0	8.590616E-05 180.5633

Exhibit 21.5 (continued)   Complex displacement vector

1.460000E+02	G	0.0	8.770063E-04	0.0	0.0	0.0	8.345494E-05
		0.0	180.0076	0.0	0.0	0.0	180.5671
1.470000E+02	G	0.0	8.575353E-04	0.0	0.0	0.0	8.111151E-05
		0.0	180.0010	0.0	0.0	0.0	180.5713
1.480000E+02	G	0.0	8.389394E-04	0.0	0.0	0.0	7.886899E-05
		0.0	179.9945	0.0	0.0	0.0	180.5757
1.490000E+02	G	0.0	8.211643E-04	0.0	0.0	0.0	7.672106E-05
		0.0	179.9880	0.0	0.0	0.0	180.5805
1.500000E+02	G	0.0	8.041598E-04	0.0	0.0	0.0	7.466183E-05
		0.0	179.9815	0.0	0.0	0.0	180.5855
1.510000E+02	G	0.0	7.878809E-04	0.0	0.0	0.0	7.268599E-05
		0.0	179.9751	0.0	0.0	0.0	180.5908
1.520000E+02	G	0.0	7.722853E-04	0.0	0.0	0.0	7.078861E-05
		0.0	179.9686	0.0	0.0	0.0	180.5964
1.530000E+02	G	0.0	7.573338E-04	0.0	0.0	0.0	6.896507E-05
		0.0	179.9622	0.0	0.0	0.0	180.6022
1.540000E+02	G	0.0	7.429908E-04	0.0	0.0	0.0	6.721120E-05
		0.0	179.9558	0.0	0.0	0.0	180.6084
1.550000E+02	G	0.0	7.292231E-04	0.0	0.0	0.0	6.552306E-05
		0.0	179.9493	0.0	0.0	0.0	180.6149
1.560000E+02	G	0.0	7.159999E-04	0.0	0.0	0.0	6.389707E-05
		0.0	179.9429	0.0	0.0	0.0	180.6216
1.570000E+02	G	0.0	7.032923E-04	0.0	0.0	0.0	6.232981E-05
		0.0	179.9364	0.0	0.0	0.0	180.6286

Exhibit 21.5 (continued)   Complex displacement vector

1.580000E+02	G	0.0	6.910741E-04	0.0	0.0	6.081818E-05
		0.0	179.9299	0.0	0.0	180.6359
1.590000E+02	G	0.0	6.793206E-04	0.0	0.0	5.935927E-05
		0.0	179.9233	0.0	0.0	180.6435
1.600000E+02	G	0.0	6.680086E-04	0.0	0.0	5.795033E-05
		0.0	179.9167	0.0	0.0	180.6514

Exhibit 21.5 (continued) Complex displacement vector

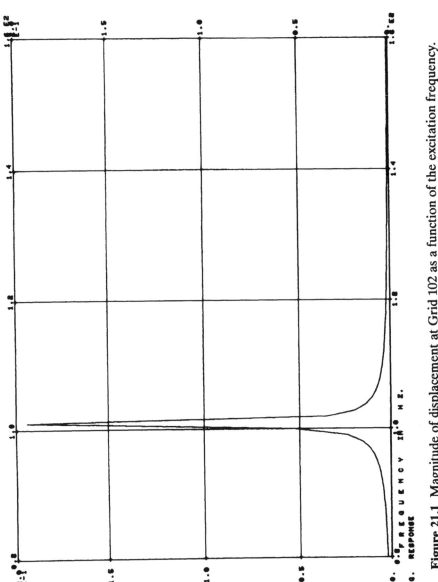

Figure 21.1 Magnitude of displacement at Grid 102 as a function of the excitation frequency.

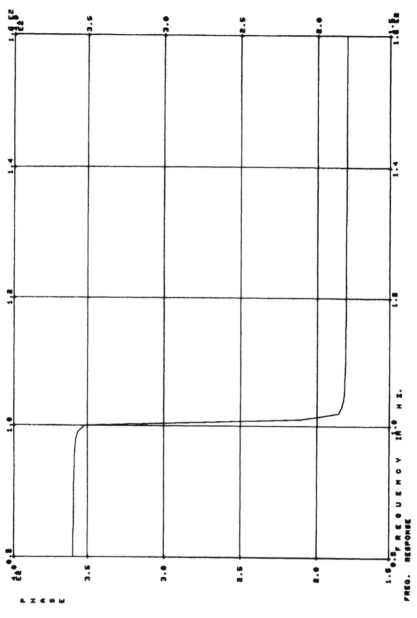

Figure 21.2 Phase angle of the response at Grid 102 as a function of the excitation frequency.

PART 3

A FEW MODELING TIPS

22
Problem 22

22.1 Statement of the problem

Consider the semicircular arch shown in Figure 22.1. Determine the reactions at the supports when a vertical load of 60 lb is applied at the center point of the arch in the negative y-direction. Assume $E = 30 \times 10^6$ psi and $\nu = 0.3333$.

22.2 Cards introduced

Case Control Deck   None

Bulk Data Deck   CORD2C

22.3 MSC/NASTRAN formulation

Since this is a symmetric structure subjected to a symmetric load, only half of it needs to be modeled. For convenience, a cylindrical coordinate system is defined to specify the GRIDs of the model. Eighteen BAR elements are used as shown in Figure 22.2. The analysis is carried out with SOL 24 (linear static analysis).

22.4 Input Data Deck

Exhibits 22.1 and 22.2 show the Executive Control Deck and the Case Control Deck, respectively.

Examine the Bulk Data Deck displayed in Exhibits 22.3 and 22.4. The cylindrical coordinate system is defined with a CORD2C card. This coordinate system, identified as 111, is employed to specify the location of the GRID points of the model. The origin of this coordinate system is at (0, 0, 0). The point (0.0, 0.0, 444.), expressed in terms of the basic (default) cartesian coordinate system, gives the orientation of the z-axis of the cylindrical coordinate system, which coincides with that of the basic coordinate system. The point (1.0, 0.0, 555.0), along with (0, 0, 0) and (0.0, 0.0, 444.), defines the plane of the azimuthal origin -- the x-z plane in this example. Thus, the GRIDs along the arch can be defined in terms of coordinates (r, θ, z) where r equals 100 in; θ varies between 0° and 90°; and z equals zero for all GRIDs.

The replicator is used to generate the GRID cards of the model. The 111 entry in the third field of the GRID cards indicates that the coordinates of the GRIDs are specified using the cylindrical coordinate system. Field 7 on the GRID cards is blank, meaning that displacement and force output at these points is expressed in terms of the basic coordinate system (x, y, z).

GRID point 444444, unlike the GRIDs on the structure, is defined using the basic coordinate system. This grid point is used only to define the orientation of Plane 1 on the CBAR cards. The CBAR cards, which are also generated with the replicator, specify that Plane 1 is the x-y plane. This means that the moment of inertia which needs to be supplied on the PBAR card is I_1. (Review Problems 7 and 9 for a definition of Plane 1 and I_1.)

The boundary conditions are defined with SPC1 cards. A symmetric boundary condition is applied at GRID 118 (only displacement in the y-direction is permitted). Note that only a half of the actual load is applied at GRID 118 since symmetry is used to analyze only a half-model.

22.5 Results

EPSILON and the reaction forces (expressed in the basic coordinate system) are shown in Exhibit 22.5. The plots are shown in Figures 22.3 and 22.4.

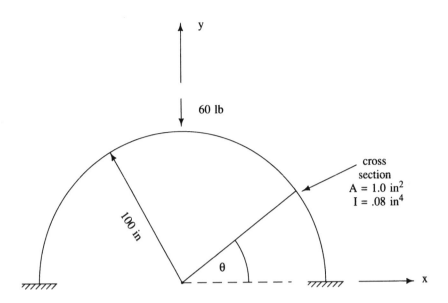

Figure 22.1

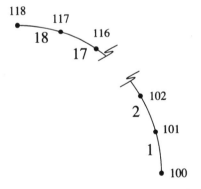

Figure 22.2

```
N A S T R A N   E X E C U T I V E   C O N T R O L   D E C K   E C H O

     ID   SSS,VVV
     SOL 24
     TIME  4
     CEND
```

Exhibit 22.1   Executive Control Deck

```
              C A S E    C O N T R O L   D E C K   E C H O
     CARD
     COUNT
       1      TITLE=    STATIC ANALYSIS USING POLAR COORDINATES
       2      $$
       3      LOAD = 3
       4      SPCFORCES= ALL
       5      ECHO=  BOTH
       6      SPC=  55
       7      $$
       8      OUTPUT(PLOT)
       9      AXES  Z,  X,  Y
      10      VIEW   0.0,  0.0,  .0
      11      FIND
      12      PLOT  LABEL   BOTH
      13      SET  44   INCLUDE  ALL
      14      PLOT  STATIC  DEFORMATION  0, SET 44
      15      BEGIN BULK
```

Exhibit 22.2   Case Control Deck

```
            I N P U T   B U L K   D A T A   D E C K   E C H O
.  1 .. 2 .. 3 .. 4 .. 5 .. 6 .. 7 .. 8 .. 9 .. 10 .
$ DEFINE CYLINDRICAL COORDINATE SYSTEM
+MMMM    1.    .0    555.          .0      .0      .0      444.      +MMMM
$    DEFINE GRID POINTS
$
GRID,100,111,100.000,0.00, .00
=,*(1),=,=, *(5.) ,=
=(17)
$    BEAMS
GRID,444444,,0.., 0.. .0.. ,123456
CBAR,1,333,100,101,444444
=,*(1),=,*(1), *(1),=
=(16)
$
PBAR,333,22,1..,.08 , ,
MAT1,22, 3.0+7, ,.3333
$
SPC1,55,123456,100
SPC1,55,345,101,THRU,117
SPC1,55,13456,118
$ FORCE
FORCE,3,118,,30.00,0.0, -1.., .0
ENDDATA
```

Exhibit 22.3   Input Bulk Data Deck

S O R T E D   B U L K   D A T A   E C H O

CARD COUNT	.	1 ..	2 ..	3 ..	4 ..	5 ..	6 ..	7 ..	8 ..	9 ..	10 .
1-	CBAR	1		333	100	101	444444				
2-	CBAR	2		333	101	102	444444				
3-	CBAR	3		333	102	103	444444				
4-	CBAR	4		333	103	104	444444				
5-	CBAR	5		333	104	105	444444				
6-	CBAR	6		333	105	106	444444				
7-	CBAR	7		333	106	107	444444				
8-	CBAR	8		333	107	108	444444				
9-	CBAR	9		333	108	109	444444				
10-	CBAR	10		333	109	110	444444				
11-	CBAR	11		333	110	111	444444				
12-	CBAR	12		333	111	112	444444				
13-	CBAR	13		333	112	113	444444				
14-	CBAR	14		333	113	114	444444				
15-	CBAR	15		333	114	115	444444				
16-	CBAR	16		333	115	116	444444				
17-	CBAR	17		333	116	117	444444				
18-	CBAR	18		333	117	118	444444				
19-	CORD2C	111		.0		.0	.0		.0		.
20-	+MMMM	1.	.0	555.		30.00	0.0	.0	444.		+MMMM
21-	FORCE	3	118	.0			-1.	.0			

Exhibit 22.4   Bulk Data Deck sorted by alphabetical order

```
22-    GRID   100      111   100.000  0.00    .00
23-    GRID   101      111   100.000  5.      .00
24-    GRID   102      111   100.000  10.     .00
25-    GRID   103      111   100.000  15.     .00
26-    GRID   104      111   100.000  20.     .00
27-    GRID   105      111   100.000  25.     .00
28-    GRID   106      111   100.000  30.     .00
29-    GRID   107      111   100.000  35.     .00
30-    GRID   108      111   100.000  40.     .00
31-    GRID   109      111   100.000  45.     .00
32-    GRID   110      111   100.000  50.     .00
33-    GRID   111      111   100.000  55.     .00
34-    GRID   112      111   100.000  60.     .00
35-    GRID   113      111   100.000  65.     .00
36-    GRID   114      111   100.000  70.     .00
37-    GRID   115      111   100.000  75.     .00
38-    GRID   116      111   100.000  80.     .00
39-    GRID   117      111   100.000  85.     .00
40-    GRID   118      111   100.000  90.     .00
41-    GRID   444444         0.       0.      .0
42-    MAT1   22       3.0+7           .3333           123456
43-    PBAR   333      22       1.     .08
44-    SPC1   55       345      101    THRU    117
45-    SPC1   55       13456    118
46-    SPC1   55       123456   100
       ENDDATA
```

Exhibit 22.4 (continued)   Bulk Data Deck sorted by alphabetical order

```
*** USER INFORMATION MESSAGE 5293 FOR DATA BLOCK KLL

LOAD SEQ. NO.        EPSILON           EXTERNAL WORK        EPSILONS LARGER THAN 0.001 ARE FLAGGED WITH ASTERISKS
     1            8.6707917E-14        4.3698182E+00


                        F O R C E S   O F   S I N G L E - P O I N T   C O N S T R A I N T

POINT ID.  TYPE       T1              T2              T3          R1          R2              R3
    100     G    -2.754633E+01   3.000000E+01     0.0          0.0         0.0         6.611877E+02
    118     G     2.754633E+01    0.0             0.0          0.0         0.0        -9.065551E+02
```

Exhibit 22.5   Epsilon and single point constraint forces

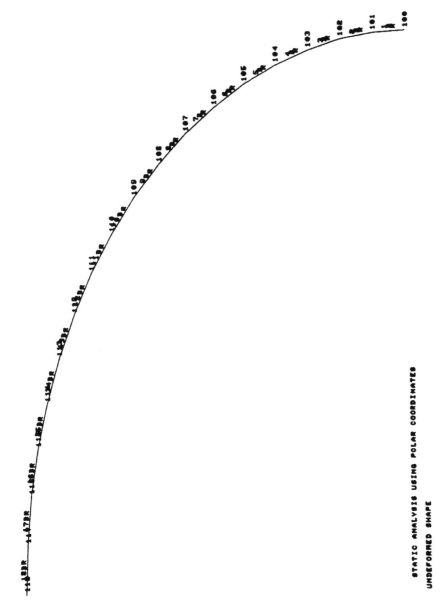

STATIC ANALYSIS USING POLAR COORDINATES

UNDEFORMED SHAPE

Figure 22.3 Finite element model of arch.

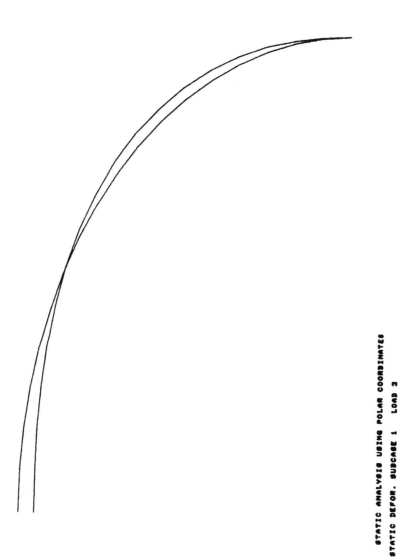

STATIC ANALYSIS USING POLAR COORDINATES

STATIC DEFOR. SUBCASE 1   LOAD 3

Figure 22.4 Deformed and undeformed shape of arch.

23

Problem 23

23.1 Statement of the problem

Consider the beam shown in Figure 23.1 ($E=3\times10^6$ psi and $\nu=0.3$). It is clamped at one end and supported by an inclined roller at the other end. Determine the response of the beam to a vertical load of 44 lb applied at its center.

23.2 Cards introduced

<u>Case Control Deck</u>   MPC

<u>Bulk Data Deck</u>   MPC

23.3 MSC/NASTRAN formulation

The structure is modeled with two BAR elements (Figure 23.2). The inclined support is modeled with an MPC (multipoint constraint) card that defines the boundary condition: at GRID 3 the displacement in the y-direction must equal the displacement in the x-direction. That is,

$$u_{3,x}-u_{3,y}=0 \tag{23.1}$$

A SOL 24 run is performed.

23.4 Input Data Deck

Exhibit 23.1 shows the Executive Control Deck. The Case Control Deck (Exhibit 23.2) includes a new card (MPC) to select the appropriate MPC card in the Bulk Data Deck.

Exhibits 23.3 and 23.4 show the Bulk Data Deck. Interpretation of the Bulk Data Deck is straightforward. The MPC card (ID=33) establishes that the displacement at GRID 3 (3 in field 3) in the x-direction (1 in field 4) multiplied by 1.00 (A=1.00 in field 5), plus the displacement at GRID 3 (3 in field 6) in the y-direction (2 in field 7) multiplied by -1.000 (A=-1.000 according to field 8) must be zero. This is exactly the condition given by Eq. 23.1.

23.5 Results

The results are shown in Exhibits 23.5 through 23.8. Note that the displacements in the x- and y-directions at GRID 3 are equal, as required. The reaction forces (forces of single point constraint) are calculated only at GRID 1. They are not computed at GRID 3 because MSC/NASTRAN does not recover multipoint constraint forces. However, they can be determined from the data shown in Exhibit 23.8 (bending moment, shear, and axial force for member 200).

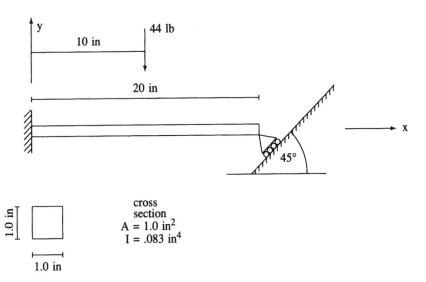

cross
section
A = 1.0 in²
I = .083 in⁴

1.0 in

1.0 in

Figure 23.1

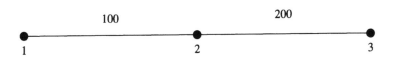

Figure 23.2

```
N A S T R A N   E X E C U T I V E   C O N T R O L   D E C K   E C H O

    ID    TTT,E
    TIME   4
    SOL 24
    CEND
```

Exhibit 23.1   Executive Control Deck

```
              C A S E    C O N T R O L   D E C K   E C H O

    CARD
    COUNT
     1     TITLE=      USE OF MPC, INCLINED PLANE
     2     LOAD = 100
     3     SPCFORCES=ALL
     4     ELFORCES  =ALL
     5     DISP = ALL
     6     $
     7     MPC= 33   $   TO SELECT  THE MPC  CARD
     8     ECHO  =BOTH
     9     BEGIN  BULK
```

Exhibit 23.2   Case Control Deck

```
            I N P U T   B U L K   D A T A   D E C K   E C H O
    .  1  ..  2  ..  3  ..  4  ..  5  ..  6  ..  7  ..  8
$
$  GRID POINTS
GRID    1                  .0      .0      .0              123456
GRID    2                10.00     .0      .0              345
GRID    3                20.0      .0      .0              345
$
$      LOAD AT POINT 2
$
FORCE   100      2                 44.     .0     -1.       .0
$
$     BAR ELEMENTS
CBAR    100      55      1       2       1.      .0      1.
CBAR    200      55      2       3       1.      .0      1.
PBAR    55       3       1.      .083    .083
$
MAT1    3       3.0+6            .3
$
$  MPC TO ENFORCE BOUNDARY CONDITION
MPC     33       3       1       1.00    3       2       -1.000
$
ENDDATA
```

Exhibit 23.3 Input Bulk Data Deck

```
            S O R T E D   B U L K   D A T A   E C H O
CARD
COUNT  .  1  ..  2  ..  3  ..  4  ..  5  ..  6  ..  7  ..  8
  ..  9  ..  10  .
  1-  CBAR    100     55      1       2       1.      .0      1.
  2-  CBAR    200     55      2       3       1.      .0      1.
  3-  FORCE   100     2               44.     .0     -1.      .0
  4-  GRID    1               .0      .0      .0              123456
  5-  GRID    2               10.00   .0      .0              345
  6-  GRID    3               20.0    .0      .0              345
  7-  MAT1    3       3.0+6           .3
  8-  MPC     33      3       1       1.00    3       2       -1.000
  9-  PBAR    55      3       1.      .083    .083
      ENDDATA
```

Exhibit 23.4 Bulk Data Deck sorted by alphabetical order

```
*** USER INFORMATION MESSAGE 5293 FOR DATA BLOCK KLL
LOAD SEQ. NO.      EPSILON         EXTERNAL WORK      EPSILONS LARGER THAN 0.001 ARE FLAGGED WITH ASTERISKS
      1          1.2025604E-17     2.8409702E-01
```

Exhibit 23.5   Epsilon

```
                                    D I S P L A C E M E N T   V E C T O R
POINT ID.  TYPE       T1              T2              T3          R1        R2          R3
    1       G        0.0             0.0             0.0         0.0       0.0         0.0
    2       G    -4.580482E-05   -1.291350E-02       0.0         0.0       0.0     -5.573619E-04
    3       G    -9.160964E-05   -9.160964E-05       0.0         0.0       0.0      2.201965E-03
```

Exhibit 23.6   Displacement vector

```
                          F O R C E S   O F   S I N G L E - P O I N T   C O N S T R A I N T
POINT ID.  TYPE       T1              T2              T3          R1        R2          R3
    1       G     1.374145E+01    3.025855E+01        0.0         0.0       0.0      1.651711E+02
```

Exhibit 23.7   Single point constraint forces

FORCES IN BAR ELEMENTS   (C B A R)

ELEMENT ID.	BEND-MOMENT END-A		BEND-MOMENT END-B		- SHEAR -		AXIAL FORCE	TORQUE
	PLANE 1	PLANE 2	PLANE 1	PLANE 2	PLANE 1	PLANE 2		
100	0.0	1.651711E+02	0.0	-1.374145E+02	0.0	3.025855E+01	-1.374145E+01	0.0
200	0.0	-1.374145E+02	0.0	2.737000E-07	0.0	-1.374145E+01	-1.374145E+01	0.0

Exhibit 23.8 Forces in BAR elements

24
Problem 24

24.1 Statement of the problem

Analyze the beam structure of Problem 23 (Figures 23.1 and 23.2) using an auxiliary coordinate system to model the inclined support.

24.2 Cards introduced

<u>Case Control Deck</u>   None

<u>Bulk Data Deck</u>   CORD1R

24.3 MSC/NASTRAN formulation

An auxiliary coordinate system is defined to measure the displacements at GRID 3. This coordinate system is such that one axis is inclined 45° with respect to the x-axis of the basic coordinate system (Figure 24.1). Therefore, to represent the support condition at GRID 3, the degree of freedom corresponding to the translation parallel to the inclined axis is left free while all other degrees of freedom except rotations about z are constrained (see Figure 24.1).

24.4 Input Data Deck

Exhibit 24.1 shows the Executive Control Deck and Exhibit 24.2 shows the Case Control Deck.

Two comments regarding the Bulk Data Deck are important (see Exhibits 24.3 and 24.4). First, an auxiliary rectangular coordinate system with its origin at GRID 3 is defined. To this end, two extra GRIDs (22 and 222) are defined. They are used only to specify the orientation of the axes of the new coordinate system; they do not serve any other purpose. This auxiliary inclined coordinate system is defined using a CORD1R card (ID = 6666). Its origin is at GRID 3, the x-axis is parallel to the inclined plane, the y-axis is perpendicular to the inclined plane, and the z-axis is parallel to the z-axis of the basic coordinate system (see Figure 24.1).

Second, the seventh field of the GRID card that defines GRID 3 states that the displacements for this point will be measured in coordinate system 6666. Consequently, only two degrees of freedom are left free in the eighth field of this card: 1 (displacement along the inclined support) and 6 (rotation about z).

24.5 Results

Exhibit 24.5 shows a small EPSILON. Exhibit 24.6 shows the displacement vector. As expected, these results are consistent with those obtained in Problem 23 (Exhibit 23.6). The displacements at GRID 2 are the same in both cases. GRID 3 exhibits the same rotation about z as in Problem 23. In addition, the displacement along the inclined support (-1.295556×10^{-4} in) agrees with the horizontal and vertical displacements obtained at GRID 3 in Problem 23:

$$-1.295556 \times 10^{-4} = -[(9.160964 \times 10^{-5})^2 + (9.160964 \times 10^{-5})^2]^{1/2} \qquad (24.1)$$

Both approaches for modeling the inclined support are, therefore, equivalent. Reaction forces and element forces are shown in Exhibits 24.7 and 24.8. They also agree with the values displayed in Exhibits 23.7 and 23.8. The value $1.943334E+01$, corresponds to the reaction at GRID 3 perpendicular to the inclined support, since the reactions at this point are expressed in terms of the auxiliary coordinate system. Recall that in Problem 23 reactions at GRID 3 were not recovered.

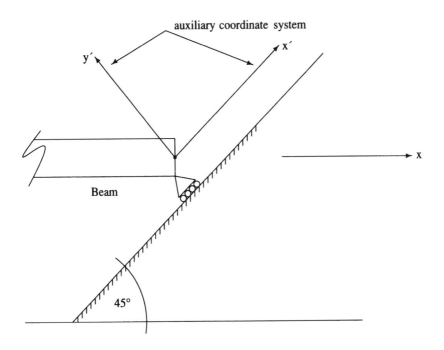

Figure 24.1

```
N A S T R A N   E X E C U T I V E   C O N T R O L   D E C K   E C H O
    ID    TTT,E
    TIME   4
    SOL 24
    CEND
```

Exhibit 24.1 Executive Control Deck

```
              C A S E   C O N T R O L   D E C K   E C H O
    CARD
    COUNT
    1       TITLE= ADDITIONAL COORDINATE SYSTEM TO ENFORCE  B.C.
    2       LOAD = 100
    3       SPCFORCES=ALL
    4       ELFORCES =ALL
    5       DISP = ALL
    6       ECHO= BOTH
    7       $
    8       BEGIN  BULK
```

Exhibit 24.2 Case Control Deck

```
            I N P U T   B U L K   D A T A   D E C K   E C H O
    .   1  ..   2  ..   3  ..   4  ..   5  ..   6  ..   7  ..   8  ..
    $  GRID POINTS
    GRID    1               .0      .0      .0                  123456
    GRID    2             10.00     .0      .0                  345
    GRID    3             20.0      .0      .0      6666        2345
    $  COORDINATE SYSTEM TO MEASURE DISPLACEMENT AT NODE  3
    CORD1R  6666     3       22      222
    GRID    22             20.0      .0    444.                 123456
    GRID    222           21.00    1.0    1.                    123456
    $      LOAD AT POINT 2
    FORCE   100     2              44.      .0     -1.      .0
    $      BAR ELEMENTS
    CBAR    100     55      1       2      1.       .0     1.
    CBAR    200     55      2       3      1.       .0     1.
    PBAR    55      3      1.       .083   .083
    MAT1    3      3.0+6            .3
    ENDDATA
```

Exhibit 24.3 Input Bulk Data Deck

```
                  S O R T E D   B U L K   D A T A   E C H O
CARD
COUNT  .  1  .. 2  .. 3  .. 4  .. 5  .. 6  .. 7  .. 8
  1-  CBAR   100   55    1     2     1.    .0    1.
  2-  CBAR   200   55    2     3     1.    .0    1.
  3-  CORD1R 6666  3     22    222
  4-  FORCE  100   2           44.   .0    -1.         .0
  5-  GRID   1           .0    .0    .0          123456
  6-  GRID   2           10.00 .0    .0          345
  7-  GRID   3           20.0  .0    .0    6666  2345
  8-  GRID   22          20.0  .0    444.        123456
  9-  GRID   222         21.00 1.0   1.          123456
 10-  MAT1   3     3.0+6       .3
 11-  PBAR   55    3     1.    .083  .083
      ENDDATA
```

Exhibit 24.4   Bulk Data Deck sorted by alphabetical order

```
*** USER INFORMATION MESSAGE 5293 FOR DATA BLOCK KLL

  LOAD SEQ. NO.        EPSILON        EXTERNAL WORK        EPSILONS LARGER THAN 0.001 ARE FLAGGED WITH ASTERISKS
      1            -1.7424919E-18      2.8409702E-01
```

Exhibit 24.5 Epsilon

| | | D I S P L A C E M E N T   V E C T O R | | | | |
POINT ID.	TYPE	T1	T2	T3	R1	R2	R3
1	G	0.0	0.0	0.0	0.0	0.0	0.0
2	G	-4.580482E-05	-1.291350E-02	0.0	0.0	0.0	-5.573619E-04
3	G	-1.295556E-04	0.0	0.0	0.0	0.0	2.201965E-03
22	G	0.0	0.0	0.0	0.0	0.0	0.0
222	G	0.0	0.0	0.0	0.0	0.0	0.0

Exhibit 24.6 Displacement vector

| | | F O R C E S   O F   S I N G L E - P O I N T   C O N S T R A I N T | | | | |
POINT ID.	TYPE	T1	T2	T3	R1	R2	R3
1	G	1.374145E+01	3.025855E+01	0.0	0.0	0.0	1.651711E+02
3	G	0.0	1.943334E+01	0.0	0.0	0.0	0.0

Exhibit 24.7 Single point constraint forces

F O R C E S   I N   B A R   E L E M E N T S   (C B A R)

ELEMENT ID.	BEND-MOMENT END-A		BEND-MOMENT END-B		- SHEAR -		AXIAL FORCE	TORQUE
	PLANE 1	PLANE 2	PLANE 1	PLANE 2	PLANE 1	PLANE 2		
100	0.0	1.651711E+02	0.0	-1.374145E+02	0.0	3.025855E+01	-1.374145E+01	0.0
200	0.0	-1.374145E+02	0.0	2.737000E-07	0.0	-1.374145E+01	-1.374145E+01	0.0

Exhibit 24.8   Forces in BAR elements

25

Problem 25

25.1 Statement of the problem

Consider the beam structure shown in Figure 25.1. Determine the reactions
and the element forces produced by a 44 lb load at the center. Assume that
the two beams are rigidly connected; therefore, both shear and bending are
transmitted. Take $E = 3 \times 10^6$ psi and $\nu = 0.3$.

25.2 Cards introduced

Case Control Deck   None

Bulk Data Deck   None

25.3 MSC/NASTRAN formulation

The structure is modeled with two BAR elements and three GRID points by
taking advantage of the offset vector feature provided on the CBAR card.
See Figure 25.2.

25.4 Input Data Deck

The Executive Control Deck and the Case Control Deck are shown in
Exhibits 25.1 and 25.2, respectively. The Bulk Data Deck (Exhibits 25.3 and
25.4) is fairly standard. Notice, however, that BAR element 200 is defined
using GRIDs 2 and 3 as reference points. An offset vector (0.0, -2.0, 0.0) is
employed to give the exact location of the left end of BAR element 200. The
components of the offset vector are defined in fields 4, 5, and 6 of the

continuation card that defines CBAR 200. The offset vector represents a rigid link between the actual end point of the BAR and GA or GB.

25.5 Results

The results are shown in Exhibits 25.5, 25.6, and 25.7.

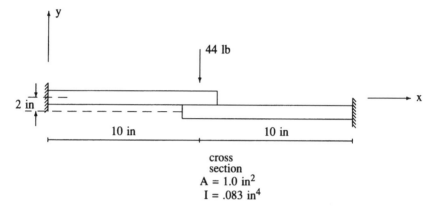

Figure 25.1

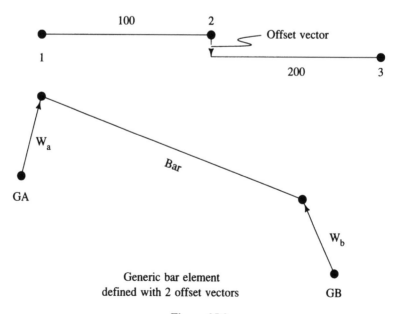

Generic bar element
defined with 2 offset vectors

Figure 25.2

```
N A S T R A N   E X E C U T I V E   C O N T R O L   D E C K   E C H O

     ID    TTT,E
     TIME  4
     SOL 24
     CEND
```

Exhibit 25.1   Executive Control Deck

```
              C A S E   C O N T R O L   D E C K   E C H O
     CARD
     COUNT
       1     TITLE= CONNECTION WITH OFFSET VECTOR
       2     LOAD = 100
       3     SPCFORCES=ALL
       4     ELFORCES =ALL
       5     $
       6     ECHO=BOTH
       7     BEGIN  BULK
```

Exhibit 25.2   Case Control Deck

```
           I N P U T   B U L K   D A T A   D E C K   E C H O

   .  1  ..  2  ..  3  ..  4  ..  5  ..  6  ..  7  ..  8  ..  9  ..  10  .
$
$  GRID POINTS
$
GRID     1                  .0       .0       .0                123456
GRID     2               10.00       .0       .0                1345
GRID     3               20.0     -2.0       .0                123456
$
$  LOAD AT POINT 2
$
FORCE    100     2                44.       .0     -1.          .0
$
$  BAR ELEMENTS
$
CBAR     100     55     1      2      1.      .0     1.
CBAR     200     55     2      3      1.      .0     1.              +BB
+BB                     .0    -2.      .0
PBAR     55      3      1.    .083   .083
$
MAT1     3      3.0+6         .3
$
$
ENDDATA
```

Exhibit 25.3 Input Bulk Data Deck

```
                              S O R T E D   B U L K   D A T A   E C H O

CARD
COUNT   .   1  ..  2  ..  3  ..  4  ..  5  ..  6  ..  7  ..  8  ..  9  ..  10  .
1-     CBAR   100    55     1      2     1.    .0    1.
2-     CBAR   200    55     2      3     1.    .0    1.                      +BB
3-     +BB                  .0    -2.    .0
4-     FORCE  100     2     .0    44.    .0   -1.    .0
5-     GRID     1            .0    .0    .0         123456
6-     GRID     2         10.00    .0    .0          1345
7-     GRID     3         20.0   -2.0    .0         123456
8-     MAT1     3  3.0+6          .3   .083
9-     PBAR    55     3     1.   .083  .083
       ENDDATA
```

Exhibit 25.4 Bulk Data Deck sorted by alphabetical order

```
*** USER INFORMATION MESSAGE 5293 FOR DATA BLOCK KLL
                                       EPSILONS LARGER THAN 0.001 ARE FLAGGED WITH ASTERISKS
    LOAD SEQ. NO.      EPSILON           EXTERNAL WORK
         1          0.0000000E+00       1.6198127E-01
```

Exhibit 25.5 Epsilon

POINT ID.	TYPE	FORCES OF SINGLE-POINT CONSTRAINT T1	T2	T3	R1	R2	R3
1	G	0.0	2.200000E+01	0.0	0.0	0.0	1.100000E+02
3	G	0.0	2.200000E+01	0.0	0.0	0.0	-1.100000E+02

Exhibit 25.6   Single point constraint forces

FORCES IN BAR ELEMENTS   (CBAR)

ELEMENT ID.	BEND-MOMENT END-A		BEND-MOMENT END-B		- SHEAR -		AXIAL FORCE	TORQUE
	PLANE 1	PLANE 2	PLANE 1	PLANE 2	PLANE 1	PLANE 2		
100	0.0	1.100000E+02	0.0	-1.100000E+02	0.0	2.200000E+01	0.0	0.0
200	0.0	-1.100000E+02	0.0	1.100000E+02	0.0	-2.200000E+01	0.0	0.0

Exhibit 25.7   Forces in BAR elements

26

Problem 26

26.1 Statement of the problem

Model the structure analyzed in Problem 25 (Figure 25.1) using a rigid element to represent the connection at the center instead of the offset vector feature of the CBAR card.

26.2 Cards introduced

Case Control Deck   None

Bulk Data Deck   RBE2

26.3 MSC/NASTRAN formulation

The structure is modeled using two BARs and four GRIDs. Each GRID point corresponds to the geometric location of the end of a BAR element (see Figure 26.1). Thus, it is necessary, somehow, to rigidly connect GRID 2 and GRID 2222 to model the actual configuration. This is done by introducing a rigid element (RBE2) between GRID 2 and GRID 2222. Internally, the RBE2 element represents a mathematical relationship linking the displacements of the connected grid points. This element does not "add stiffness" to the stiffness matrix of the structure. Therefore, it does not cause numerical conditioning problems.

26.4 Input Data Deck

Exhibits 26.1 and 26.2 show the Executive Control Deck and the Case Control Deck respectively. It can be seen that an extra GRID (2222) is included in the Bulk Data Deck (Exhibit 26.3) to represent the left end of BAR 200. In addition, a rigid bar is defined with an RBE2 card.

The third field of the RBE2 card (2222) represents the "master" node, the fourth field (26) specifies the degrees of freedom that will be rigidly connected (y-displacement and rotation about z), and field 5 gives the ID number of the "slave" GRID (2). In this particular case, either node could have been chosen as the "master" node. Additional "slave" GRIDs could also be incorporated on the RBE2 card, if desired. In summary, the RBE2 card states that the displacement at GRID 2 in the y-direction will be equal to the displacement at GRID 2222 in the y-direction; and the rotation about z for GRID 2 will be equal to the rotation about z for GRID 2222.

26.5 Results

The results are shown in Exhibits 26.5, 26.6, and 26.7. They coincide with the results obtained in Problem 25 (Exhibits 25.5, 25.6 and 25.7).

26.6 Additional Comments

Both approaches (the use of the offset vector and the rigid element) are equally valid to model the type of connection presented in Problem 25 (Figure 25.1). The RBE2, however, represents a more general technique since it is not restricted to BAR elements (it operates at the GRID level). Thus, the RBE2 element can also be used to model any type of rigid link connecting two or more arbitrary points of the model, and involving some or all six degrees of freedom.

Finally, it is not a good modeling practice to represent rigid links using very stiff BARs. For instance, it might have been tempting to define a BAR with a very large moment of inertia between GRIDs 2 and 2222. This approach, however, is likely to produce ill-conditioned stiffness matrices resulting in unreliable results.

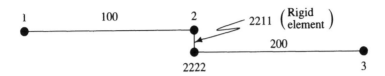

Figure 26.1

```
N A S T R A N   E X E C U T I V E   C O N T R O L   D E C K   E C H O
     ID    TTT,E
     TIME   4
     SOL 24
     CEND
```

Exhibit 26.1   Executive Control Deck

```
            C A S E   C O N T R O L   D E C K   E C H O
  CARD
  COUNT
    1       TITLE= RIGID  ELEMENT TO CONNECT THE 2 BEAMS
    2       LOAD = 100
    3       SPCFORCES=ALL
    4       ELFORCES=ALL
    5       ECHO=BOTH
    6       $
    7       BEGIN  BULK
```

Exhibit 26.2   Case Control Deck

```
                I N P U T    B U L K    D A T A    D E C K    E C H O

      .  1  ..  2  ..  3  ..  4  ..  5  ..  6  ..  7  ..  8  ..
      $
      $  GRID POINTS
      $
      GRID    1                .0      .0      .0                  123456
      GRID    2             10.00      .0      .0                  1345
      GRID    2222          10.00    -2.0      .0                  1345
      GRID    3             20.0     -2.0      .0                  123456
      $
      $       R I G I D    E L E M E N T
      $-----------------------------------
      RBE2,2211,2222,26,2
      $-----------------------------------
      $  THIS CARD MEANS THAT A) GRID POINT 2222 IS THE "MASTER"  NODE
      $                        B) GRID POINT 2 IS THE "SLAVE"  NODE
      $                   AND  C) A RIGID ELEMENT CONNECTS DEGREES
      $                           OF FREEDOM 2  AND  6 BETWEEN THE
      $                           MASTER NODE AND THE SLAVE NODE.
      $
      $     LOAD AT POINT 2
      $
      FORCE   100     2                44.     .0     -1.     .0
      $
      $    BAR ELEMENTS
      $
      CBAR    100     55      1       2      1.      .0     1.
      CBAR    200     55      3       2222   1.      .0     1.
      PBAR    55      3       1.      .083   .083
      $
      MAT1    3       3.0+6           .3
      $
      $
      $
      ENDDATA
```

Exhibit 26.3   Input Bulk Data Deck

```
            S O R T E D   B U L K   D A T A   E C H O

CARD
COUNT  .  1  ..  2  ..  3  ..  4  ..  5  ..  6  ..  7  ..  8
  1-  CBAR    100    55    1       2      1.    .0    1.
  2-  CBAR    200    55    3       2222   1.    .0    1.
  3-  FORCE   100    2             44.    .0   -1.    .0
  4-  GRID    1             .0     .0     .0          123456
  5-  GRID    2             10.00  .0     .0          1345
  6-  GRID    3             20.0   -2.0   .0          123456
  7-  GRID    2222          10.00  -2.0   .0          1345
  8-  MAT1    3      3.0+6          .3
  9-  PBAR    55     3       1.     .083   .083
 10-  RBE2    2211   2222    26     2
      ENDDATA
```

Exhibit 26.4   Bulk Data Deck sorted by alphabetical order

```
*** USER INFORMATION MESSAGE 5293 FOR DATA BLOCK KLL
   LOAD SEQ. NO.      EPSILON          EXTERNAL WORK          EPSILONS LARGER THAN 0.001 ARE FLAGGED WITH ASTERISKS
        1          0.0000000E+00       1.6198127E-01
```

Exhibit 26.5 Epsilon

```
                    F O R C E S   O F   S I N G L E - P O I N T   C O N S T R A I N T
   POINT ID.  TYPE     T1            T2            T3           R1          R2            R3
       1       G      0.0       2.200000E+01      0.0          0.0         0.0       1.100000E+02
       3       G      0.0       2.200000E+01      0.0          0.0         0.0      -1.100000E+02
```

Exhibit 26.6 Single point constraint forces

```
                    F O R C E S   I N   B A R   E L E M E N T S     ( C B A R )
ELEMENT       BEND-MOMENT END-A       BEND-MOMENT END-B        - SHEAR -            AXIAL
   ID.     PLANE 1     PLANE 2     PLANE 1      PLANE 2     PLANE 1     PLANE 2      FORCE      TORQUE
   100      0.0      1.100000E+02    0.0    -1.100000E+02    0.0     2.200000E+01    0.0        0.0
   200      0.0     -1.100000E+02    0.0     1.100000E+02    0.0    -2.200000E+01    0.0        0.0
```

Exhibit 26.7 Forces in BAR elements

27

Problem 27

27.1 Statement of the problem

Consider the simple beam structure shown in Figure 27.1. The properties of the beam are as follows: $E = 3 \times 10^6$ psi, $\nu = .3333$, and ρ (mass per unit of volume) $= 7 \times 10^{-4}$ lb-s^2/in^4. The beam is excited by a base acceleration f(t) in the y-direction given in units of in/s^2 by

$$f(t) = -e^{-t}\cos(2\pi t) \qquad (27.1)$$

Determine the forces in the structure and the displacement at the top (relative to the base) produced by this excitation. This is a typical base-excitation problem, also called an enforced acceleration problem. This class of problems is of considerable interest in earthquake engineering.

27.2 Cards introduced

<u>Case Control Deck</u>   None

<u>Bulk Data Deck</u>   None

27.3 MSC/NASTRAN formulation

The structure is modeled with two BAR elements as depicted in Figure 27.2. This example presents one possibility for representing the effect of base acceleration. The equation of motion of the system can be written as

$$[M](d^2\{y\}/dt^2 + f(t)) + [K]\{y\} = \{0\} \qquad (27.2)$$

where [M] is the mass matrix; [K] is the stiffness matrix; f(t) is the base acceleration; and {y} is the relative displacement of the structure with respect to its base. Eq 27.2 can be rewritten as

$$[M]d^2\{y\}/dt^2 + [K]\{y\} = -[M]\{f(t)\} \qquad (27.3)$$

Eq. 27.3 shows that the problem can be analyzed simply by exciting the structure at each node with a force equal to the mass lumped at that node multiplied by the base acceleration (with a negative sign). In this case, the mass m_2 lumped at GRID 2 is given by

$$m_2 = \rho Al = 0.0007 \times 50 \times 75 = 2.625 \text{ lb-s}^2/\text{in} \qquad (27.4)$$

where A represents the area of the cross section of the BAR and l is its length. The mass lumped at GRID 3 is half this value.

27.4 Input Data Deck

SOL 27 (direct transient response) is selected in the Executive Control Deck (Exhibit 27.1). The Case Control Deck is shown in Exhibit 27.2.

A few comments regarding the Bulk Data Deck (Exhibits 27.3 and 27.4) are necessary. The forces acting on the structure are defined with a DLOAD card (selected by DLOAD=2222 in the Case Control Deck). This card combines two TLOAD2 cards which in turn point to two DAREA cards. The force acting at a GRID point, according to Eq. 27.3, is the product of two factors: the mass concentrated at that GRID and the function -f(t). This is consistent with what is observed in the Bulk Data Deck. The TLOAD2 card is actually used to define the function -f(t) while the DAREA card is used to define a scaling factor equal to the mass concentrated at the point where the load is applied.

The TSTEP card specifies a time step equal to .01 s. This is adequate since the natural period of this structure is equal to 0.075 s, about seven times Δt.

27.5 Results

Exhibit 27.5 shows the displacement at GRID 3 as a function of time. Recall that in this context T2 represents the relative displacement with respect to the foundation. Exhibit 27.6 shows the bending moment and shear as a function of time for BAR 10.

27.6 Additional comments

The use of the GRAV card in dynamics analyses (see Problem 19) provides an alternative way to model base acceleration effects. In this case, the GRAV card is used to define a "gravity acceleration" equal to one in the desired direction. A TABLED1 card is used to specify the time history of the base acceleration. In essence, this procedure consists of defining a "time dependent gravity load" equal to the base acceleration.

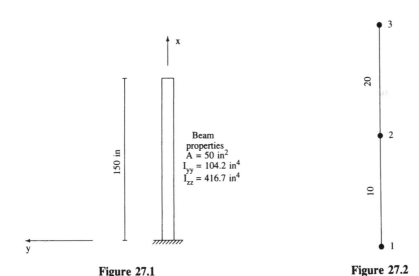

Beam properties
$A = 50$ in^2
$I_{yy} = 104.2$ in^4
$I_{zz} = 416.7$ in^4

Figure 27.1   **Figure 27.2**

```
NASTRAN  EXECUTIVE  CONTROL  DECK  ECHO



    ID   XXX,XXX
    SOL  27
    TIME  5
    CEND
```

Exhibit 27.1   Executive Control Deck

```
            C A S E   C O N T R O L   D E C K   E C H O
    CARD
    COUNT
     1      TITLE= ANALYSIS OF BEAM---- TRANSIENT RESPONSE
     2      SET  33=   3
     3      DISPLACEMENT=33
     4      SET  444=10
     5      ELFORCES =444
     6      $
     7      TSTEP= 99
     8      DLOAD= 2222
     9      $
    10      ECHO=  BOTH
    11      $
    12      $
    13      BEGIN BULK
```

Exhibit 27.2   Case Control Deck

```
                    I N P U T   B U L K   D A T A   D E C K   E C H O

    .  1  ..  2  ..  3  ..  4  ..  5  ..  6  ..  7  ..  8  ..  9  ..  10  .
    $
    $
    $  GRID  POINTS
    $
    GRID    1                 .0        .0        .0              123456
    GRID    2              75.00        .0        .0              1345
    GRID    3              150.0        .0        .0              1345
    $
    $
    CBAR    10    3333       1         2         1.        .0      1.
    CBAR    20    3333       2         3         1.        0.      1.
    $
    $
    PBAR    3333  1212    50.00    104.20    416.70
    $
    $
    MAT1    1212  30.+6     .333      7.-4
    $
    $
```

Exhibit 27.3   Input Bulk Data Deck

```
                I N P U T   B U L K   D A T A   D E C K   E C H O

     .   1  ..  2  ..  3  ..  4  ..  5  ..  6  ..  7  ..  8  ..  9  ..  10  .
     $
     $ FORCE ACTING AT GRID 2 AND 3
     $
     $
     DLOAD   2222   1.000   1.000   100     1.000   200
     TLOAD2  100    101     .0              0.000   10.+8   1.00    .0      +LK
     +LK     -1.    .0
     TLOAD2  200    201     .0              0.000   10.+8   1.00    .0      +MMM
     +MMM    -1.    .0
     $
     $
     DAREA   101    2       2       2.625
     DAREA   201    3       2       1.3125
     $
     $
     TSTEP   99     100     .01     1
     ENDDATA
```

Exhibit 27.3 (continued) Input Bulk Data Deck

S O R T E D   B U L K   D A T A   E C H O

CARD COUNT	. 1	.. 2	.. 3	.. 4	.. 5	.. 6	.. 7	.. 8	.. 9	.. 10 .
1-	CBAR	10	3333	1	2	1.	.0	1.		
2-	CBAR	20	3333	2	3	1.	0.	1.		
3-	DAREA	101	2	2	2.625					
4-	DAREA	201	3	2	1.3125					
5-	DLOAD	2222	1.000	1.000	100	1.000	200			
6-	GRID	1	.0	.0	.0	.0		123456		
7-	GRID	2	75.00	.0	.0	.0		1345		
8-	GRID	3	150.0	.0	.0	.0		1345		
9-	MAT1	1212	30.+6		.333	7.-4				
10-	PBAR	3333	1212	50.00	104.20	416.70				
11-	TLOAD2	100	101			0.000	10.+8	1.00	.0	+LK
12-	+LK	-1.	.0							
13-	TLOAD2	200	201			0.000	10.+8	1.00	.0	+MMM
14-	+MMM	-1.	.0							
15-	TSTEP	99	100	.01	1					
	ENDDATA									

Exhibit 27.4   Bulk Data Deck sorted by alphabetical order

POINT-ID =   3

| | | | D I S P L A C E M E N T   V E C T O R | | | | |
TIME	TYPE	T1	T2	T3	R1	R2	R3
0.0		0.0	0.0	0.0	0.0	0.0	0.0
1.000000E-02	G	0.0	3.470723E-05	0.0	0.0	0.0	2.651091E-07
2.000000E-02	G	0.0	1.218519E-04	0.0	0.0	0.0	1.078890E-06
3.000000E-02	G	0.0	2.466289E-04	0.0	0.0	0.0	2.303144E-06
4.000000E-02	G	0.0	3.350706E-04	0.0	0.0	0.0	3.178638E-06
5.000000E-02	G	0.0	3.334309E-04	0.0	0.0	0.0	3.136401E-06
6.000000E-02	G	0.0	2.411713E-04	0.0	0.0	0.0	2.244960E-06
7.000000E-02	G	0.0	1.077815E-04	0.0	0.0	0.0	9.890548E-07
8.000000E-02	G	0.0	4.301153E-06	0.0	0.0	0.0	-2.270565E-08
9.000000E-02	G	0.0	-1.342201E-05	0.0	0.0	0.0	-2.121372E-07

S O M E   L I N E S   H A V E   B E E N   D E L E T E D

TIME	TYPE	T1	T2	T3	R1	R2	R3
9.099994E-01	G	0.0	-3.610608E-05	0.0	0.0	0.0	-3.616605E-07
9.199994E-01	G	0.0	8.933426E-05	0.0	0.0	0.0	8.409060E-07
9.299994E-01	G	0.0	2.016995E-04	0.0	0.0	0.0	1.899957E-06
9.399994E-01	G	0.0	2.390817E-04	0.0	0.0	0.0	2.284090E-06
9.499994E-01	G	0.0	1.797537E-04	0.0	0.0	0.0	1.722631E-06
9.599994E-01	G	0.0	5.743099E-05	0.0	0.0	0.0	5.101459E-07
9.699994E-01	G	0.0	-5.644632E-05	0.0	0.0	0.0	-5.842589E-07
9.799994E-01	G	0.0	-9.714422E-05	0.0	0.0	0.0	-9.502455E-07
9.899994E-01	G	0.0	-4.278015E-05	0.0	0.0	0.0	-4.466281E-07
9.999993E-01	G	0.0	7.655642E-05	0.0	0.0	0.0	6.919448E-07

Exhibit 27.5   Displacement vector as a function of time

ELEMENT-ID =   10

F O R C E S   I N   B A R   E L E M E N T S   (C B A R)

| TIME | BEND-MOMENT-END-A | | BEND-MOMENT-END-B | | SHEAR | | FORCE | TORQUE |
	PLANE 1	PLANE 2	PLANE 1	PLANE 2	PLANE 1	PLANE 2		
0.0	0.0	0.0	0.0	0.0	0.0	0.0	0.0	0.0
1.000000E-02	0.0	-9.883489E+01	0.0	5.229068E+00	0.0	-1.387519E+00	0.0	0.0
2.000000E-02	0.0	-2.729232E+02	0.0	-4.336781E+01	0.0	-3.060738E+00	0.0	0.0
3.000000E-02	0.0	-4.926603E+02	0.0	-1.375579E+02	0.0	-4.734699E+00	0.0	0.0
4.000000E-02	0.0	-6.445367E+02	0.0	-2.075470E+02	0.0	-5.826530E+00	0.0	0.0
5.000000E-02	0.0	-6.547242E+02	0.0	-1.954132E+02	0.0	-6.124146E+00	0.0	0.0
6.000000E-02	0.0	-4.853670E+02	0.0	-1.315065E+02	0.0	-4.718141E+00	0.0	0.0
7.000000E-02	0.0	-2.240340E+02	0.0	-5.283867E+01	0.0	-2.282604E+00	0.0	0.0
8.000000E-02	0.0	-4.003038E+01	0.0	2.379977E+01	0.0	-8.510687E-01	0.0	0.0
9.000000E-02	0.0	-1.658985E+01	0.0	4.365396E+01	0.0	-8.032507E-01	0.0	0.0
		S O M E L I N E S H A V E B E E N D E L E T E D						
9.099994E-01	0.0	5.988174E+01	0.0	3.034069E+01	0.0	3.938807E-01	0.0	0.0
9.199994E-01	0.0	-1.751228E+02	0.0	-5.260083E+01	0.0	-1.633626E+00	0.0	0.0
9.299994E-01	0.0	-3.947162E+02	0.0	-1.193268E+02	0.0	-3.671858E+00	0.0	0.0
9.399994E-01	0.0	-4.518686E+02	0.0	-1.547780E+02	0.0	-3.961208E+00	0.0	0.0
9.499994E-01	0.0	-3.370696E+02	0.0	-1.185934E+02	0.0	-2.913016E+00	0.0	0.0
9.599994E-01	0.0	-1.278106E+02	0.0	-2.112582E+01	0.0	-1.422464E+00	0.0	0.0
9.699994E-01	0.0	8.418610E+01	0.0	5.529124E+01	0.0	3.852648E-01	0.0	0.0
9.799994E-01	0.0	1.725192E+02	0.0	7.212731E+01	0.0	1.338559E+00	0.0	0.0
9.899994E-01	0.0	6.189185E+01	0.0	4.349806E+01	0.0	2.452505E-01	0.0	0.0
9.999993E-01	0.0	-1.644169E+02	0.0	-3.312494E+01	0.0	-1.750559E+00	0.0	0.0

Exhibit 27.6 Forces in BAR elements as a function of time

28
Problem 28

28.1 Statement of the problem

Determine the response of the beam structure presented in Problem 27
(Figure 27.1) using the so-called big mass approach.

28.2 Cards introduced

Case Control Deck   None

Bulk Data Deck   None

28.3 MSC/NASTRAN formulation

The same model as Problem 27 is used (see Figures 27.1 and 27.2). The
acceleration is applied directly at the base of the structure using a numerical
trick. A big mass, say M, is attached at the base of the structure. Then, a
force equal to $M \times f(t)$, where $f(t)$ is the given acceleration, is applied to the
big mass. Since

$$Acceleration = Force/Mass$$

it turns out that the acceleration of the whole system (structure plus big
mass) is

$$(M \times f(t))/[M + m]$$

where m is the mass of the structure. Obviously, since M is much larger than
m, the above mentioned expression is very close to $f(t)$ and therefore the
acceleration acting on the structure is a very good approximation to $f(t)$. This

approach is necessary because MSC/NASTRAN does not provide a "direct" way to enforce a given base acceleration on a structure.

With this formulation, the displacements determined in the analysis are absolute displacements. Since we want to determine the displacement of the structure with respect to its base, an extra GRID point needs to be defined to carry this information. Finally, using a multipoint constraint (MPC) relationship, the "displacement" associated with this extra GRID point is defined as the relative displacement of the GRID in question with respect to the base.

28.4 Input Data Deck

SOL 27 (direct transient response) is selected in the Executive Control Deck (Exhibit 28.1). The Case Control Deck (Exhibit 28.2) is similar to that of Problem 27 except for the MPC=55 statement included to select an MPC card from the Bulk Data Deck.

The Bulk Data Deck (Exhibits 28.3 and 28.4) reveals a few changes compared to Problem 27. First, note the extra GRID (30) with one degree of freedom free (2). The MPC card states that the displacement at GRID 30 in the y-direction is equal to the displacement at GRID 3 minus the displacement at GRID 1. Therefore, GRID 30 represents the relative displacement at the top (GRID 3) with respect to the base (GRID 1). Second, notice that the card that defines GRID 1 (the base) shows that degree of freedom 2 (y-displacement) is free. This is consistent with the fact that the ground acceleration is applied at this point in the y-direction. Third, a big mass is attached to the base of the structure to provide inertial effects in the y-direction. This mass is defined using a CMASS2 card. The value of 10,000 lb-s^2/in is reasonably large compared to the total mass of the beam structure (5.25 lb-s^2/in). In general, a value between 10^3 to 10^6 times the total mass of the structure is adequate.

The force acting at GRID 1 is defined by the TLOAD2 card which, in turn, points to a DAREA card. The TLOAD2 card defines the ground acceleration (f(t)). The DAREA card includes a scale factor equal to the big mass, consistent with the approach previously outlined.

28.5 Results

Exhibit 28.5 shows the value of the displacement at GRID 30 as a function of time. Exhibit 28.6 shows the bending moment and shear for BAR element 10. These results are very similar to the results from Problem 27 (Exhibits 27.5 and 27.6). The differences are due to the fact that the big mass approach is only an approximation; theoretically, an infinite mass would be required for a "perfect" model. However, an excessively large value for M would not necessarily improve the accuracy of the results; it would most likely produce an ill-conditioned problem.

28.6 Additional comments

This example, and the one illustrated in Problem 27, show two different techniques to simulate the effect of ground shaking on a structure. These techniques are useful in the analysis of complex structures subjected to earthquakes. Both approaches are equally valid. However, the big mass method is easier to use since it is not necessary to estimate the mass lumped at each GRID.

Sometimes, simply for numerical stability, it is advisable to attach the big mass to a grounded spring and a grounded dashpot to prevent the big mass from "drifting away" causing numerical difficulties (Figure 28.1). This trick can make the problem more stable from a numerical standpoint without altering the response of the system. Based on previous numerical experience, it is recommended to take a spring constant of the order $\omega^2 m$ and a dashpot constant of the order of $2\omega m \zeta$ where, m is the mass of the structure, ω is the fundamental frequency of the structure, and ζ is the fraction of critical damping which for these cases must be approximately 60%.

The fundamental frequency of this structure is 13.3 Hz (T=0.075 s) and the total mass is

$$m = .0007 \times 50 \times 150 = 5.25 \text{ lb-s}^2/\text{in}$$

therefore,

$$\omega^2 m = (2\pi 13.3)^2 \times 5.25 = 36,662 \text{ lb/in} \tag{28.1}$$

and

$$2\omega m \zeta = 2 \times (2\pi 13.3) \times 5.25 \times 0.6 = 526 \text{ lb-s/in} \tag{28.2}$$

Exhibit 28.7 presents the Bulk Data Deck for a run that demonstrates this approach (big mass plus additional spring and dashpot). A spring with a constant of 40,000 lb/in and a dashpot with a constant equal to 500 lb-s/in were connected to the big mass. The results, as it can be seen in Exhibit 28.8, are almost identical to the results obtained without the spring and dashpot. In a more complex configuration, however, the auxiliary spring and dashpot could be crucial in keeping the numerical integration scheme stable.

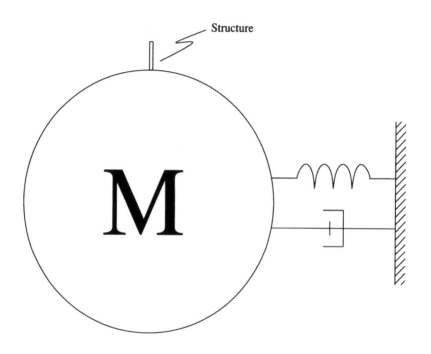

Figure 28.1

```
N A S T R A N   E X E C U T I V E   C O N T R O L   D E C K   E C H O

    ID   YYY,YYY
    SOL  27
    TIME  5
    CEND
```

Exhibit 28.1   Executive Control Deck

```
              C A S E    C O N T R O L    D E C K    E C H O
CARD
COUNT
   1      TITLE= ANALYSIS OF BEAM---- TRANSIENT RESPONSE
   2      SUBTITLE= BASE ACCELERATION
   3      SET  33=   30
   4.     DISPLACEMENT=33
   5      SET   555=10
   6      ELFORCES=555
   7      $
   8      TSTEP= 99
   9      DLOAD= 2222
  10      $
  11      ECHO=  BOTH
  12      $
  13      $
  14      MPC=55  $$$$$$$$$$ TO SELECT  THE  MPC CARD
  15      $
  16      BEGIN BULK
```

Exhibit 28.2   Case Control Deck

```
                    I N P U T   B U L K   D A T A   D E C K   E C H O

     .   1  ..   2  ..   3  ..   4  ..   5  ..   6  ..   7  ..   8  ..   9  ..  10  .
$
MPC     55      30      2       1.      3       2       -1.000
+MMMM           1       2       1.00                                            +MMMM
$
$ GRID 30 WILL REPRESENT THE RELATIVE DISPL.  OF
$ GRID 3 WITH RESPECT TO THE BASE ( GRID 1 )
$
GRID    30              555.    555.    555.                    13456
$
$
$ GRID POINTS
$
GRID    1               .0      .0      .0                      13456
GRID    2       75.00   .0      .0      .0                      1345
GRID    3       150.0   .0      .0      .0                      1345
$
$
CBAR    10      3333    1       2       1.      .0      1.
CBAR    20      3333    2       3       1.      0.      1.
$
PBAR    3333    1212    50.00   104.20  416.70
```

Exhibit 28.3   Input Bulk Data Deck

```
            I N P U T   B U L K   D A T A   D E C K   E C H O

    .  1  ..  2  ..  3  ..  4  ..  5  ..  6  ..  7  ..  8  ..  9  .. 10  .
   $
   MAT1    1212    30.+6          .333    7.-4
   $
   $
   TLOAD2  2222    101            3       0.000   10.+8   1.00    .0      +LK
   +LK     -1.     .0
   $
   $ THE "SCALING FACTOR"  10.+5 IS DETERMINED BY
   $ THE VALUE OF THE BIG MASS
   $
   DAREA   101     1       2      -10.+5
   $
   $ BIG MASS  AT THE BASE
   $
   CMASS2  3333    10.+5   1       2
   $
   TSTEP   99      100     .01     1
   ENDDATA
```

Exhibit 28.3 (continued) Input Bulk Data Deck

```
                              S O R T E D   B U L K   D A T A   E C H O

CARD
COUNT   .      1  ..   2   ..   3   ..   4    ..   5   ..  6   ..  7  ..   8   ..  9  ..  10 .
 1-    CBAR    10    3333      1       2                 1.     .0    1.
 2-    CBAR    20    3333      2       3                 1.     0.    1.
 3-    CMASS2  3333  10.+5     1       2
 4-    DAREA   101   1         2     -10.+5
 5-    GRID    1               .0      .0      .0                     13456
 6-    GRID    2            75.00      .0      .0                     1345
 7-    GRID    3           150.0       .0      .0                     1345
 8-    GRID    30          555.      555.    555.                     13456
 9-    MAT1    1212  30.+6           .333    7.-4
10-    MPC     55    30        2      1.       3       2           -1.000
11-    +MMMM         1         2                                  +MMMM
12-    PBAR    3333  1212   50.00    1.00   416.70
13-    TLOAD2  2222  101              3       3     0.000  10.+8  1.00   .0  +LK
14-    +LK     -1.   .0
15-    TSTEP   99    100      .01      1
       ENDDATA
```

Exhibit 28.4   Bulk Data Deck sorted by alphabetical order

POINT-ID =   30

TIME	TYPE	T1	DISPLACEMENT VECTOR					
			T2	T3	R1	R2	R3	
0.0	G	0.0	0.0	0.0	0.0	0.0	0.0	
1.000000E-02	G	0.0	3.470713E-05	0.0	0.0	0.0	0.0	
2.000000E-02	G	0.0	1.218516E-04	0.0	0.0	0.0	0.0	
3.000000E-02	G	0.0	2.466280E-04	0.0	0.0	0.0	0.0	
4.000000E-02	G	0.0	3.350690E-04	0.0	0.0	0.0	0.0	
5.000000E-02	G	0.0	3.334288E-04	0.0	0.0	0.0	0.0	
6.000000E-02	G	0.0	2.411691E-04	0.0	0.0	0.0	0.0	
7.000000E-02	G	0.0	1.077800E-04	0.0	0.0	0.0	0.0	
8.000000E-02	G	0.0	4.300630E-06	0.0	0.0	0.0	0.0	
9.000000E-02	G	0.0	-1.342146E-05	0.0	0.0	0.0	0.0	

SOME LINES HAVE BEEN DELETED

TIME	TYPE	T1	T2	T3	R1	R2	R3
9.099994E-01	G	0.0	-3.609386E-05	0.0	0.0	0.0	0.0
9.199994E-01	G	0.0	8.934882E-05	0.0	0.0	0.0	0.0
9.299994E-01	G	0.0	2.017079E-04	0.0	0.0	0.0	0.0
9.399994E-01	G	0.0	2.390787E-04	0.0	0.0	0.0	0.0
9.499994E-01	G	0.0	1.797407E-04	0.0	0.0	0.0	0.0
9.599994E-01	G	0.0	5.741491E-05	0.0	0.0	0.0	0.0
9.699994E-01	G	0.0	-5.645647E-05	0.0	0.0	0.0	0.0
9.799994E-01	G	0.0	-9.714270E-05	0.0	0.0	0.0	0.0
9.899994E-01	G	0.0	-4.276781E-05	0.0	0.0	0.0	0.0
9.999993E-01	G	0.0	7.657250E-05	0.0	0.0	0.0	0.0

Exhibit 28.5   Displacement vector as a function of time

ELEMENT-ID =   10

F O R C E S   I N   B A R   E L E M E N T S   (C B A R)

TIME	BEND-MOMENT-END-A		BEND-MOMENT-END-B		SHEAR		FORCE	TORQUE
	PLANE 1	PLANE 2	PLANE 1	PLANE 2	PLANE 1	PLANE 2		
0.0	0.0	0.0	0.0	0.0	0.0	0.0	0.0	0.0
1.000000E-02	0.0	-9.883463E+01	0.0	5.229053E+00	0.0	-1.387516E+00	0.0	0.0
2.000000E-02	0.0	-2.729223E+02	0.0	-4.336769E+01	0.0	-3.060728E+00	0.0	0.0
3.000000E-02	0.0	-4.926583E+02	0.0	-1.375574E+02	0.0	-4.734678E+00	0.0	0.0
4.000000E-02	0.0	-6.445336E+02	0.0	-2.075461E+02	0.0	-5.826499E+00	0.0	0.0
5.000000E-02	0.0	-6.547201E+02	0.0	-1.954121E+02	0.0	-6.124107E+00	0.0	0.0
6.000000E-02	0.0	-4.853629E+02	0.0	-1.315052E+02	0.0	-4.718102E+00	0.0	0.0
7.000000E-02	0.0	-2.240310E+02	0.0	-5.283770E+01	0.0	-2.282577E+00	0.0	0.0
8.000000E-02	0.0	-4.002936E+01	0.0	2.380008E+01	0.0	-8.510592E-01	0.0	0.0
9.000000E-02	0.0	-1.659078E+01	0.0	4.365350E+01	0.0	-8.032570E-01	0.0	0.0
			S O M E L I N E S H A V E B E E N D E L E T E D					
9.099994E-01	0.0	5.985939E+01	0.0	3.033218E+01	0.0	3.936962E-01	0.0	0.0
9.199994E-01	0.0	-1.751496E+02	0.0	-5.261084E+01	0.0	-1.633850E+00	0.0	0.0
9.299994E-01	0.0	-3.947314E+02	0.0	-1.193327E+02	0.0	-3.671982E+00	0.0	0.0
9.399994E-01	0.0	-4.518628E+02	0.0	-1.547762E+02	0.0	-3.961155E+00	0.0	0.0
9.499994E-01	0.0	-3.370456E+02	0.0	-1.185844E+02	0.0	-2.912815E+00	0.0	0.0
9.599994E-01	0.0	-1.277811E+02	0.0	-2.111468E+01	0.0	-1.422219E+00	0.0	0.0
9.699994E-01	0.0	8.420488E+01	0.0	5.529811E+01	0.0	3.854235E-01	0.0	0.0
9.799994E-01	0.0	1.725164E+02	0.0	7.212628E+01	0.0	1.338535E+00	0.0	0.0
9.899994E-01	0.0	6.186903E+01	0.0	4.348966E+01	0.0	2.450582E-01	0.0	0.0
9.999993E-01	0.0	-1.644462E+02	0.0	-3.313622E+01	0.0	-1.750800E+00	0.0	0.0

Exhibit 28.6   Forces in BAR elements as a function of time

```
              I N P U T   B U L K   D A T A   D E C K   E C H O
  .  1  ..  2  ..  3  ..  4  ..  5  ..  6  ..  7  ..  8  ..  9  ..  10  .
$
MPC      55      30      2      1.      3      2     -1.000
+MMMM            1       2      1.00                                        +MMMM
$
$ GRID 30 WILL REPRESENT THE RELATIVE DISPL. OF
$ GRID 3 WITH RESPECT TO THE BASE ( GRID 1 )
$
GRID     30             555.   555.   555.                13456
$
$ GRID POINTS
$
GRID     1              .0     .0     .0                  13456
GRID     2              75.00  .0     .0                  1345
GRID     3              150.0  .0     .0                  1345
$
CBAR     10     3333    1      2      1.      1.     .0    1.
CBAR     20     3333    2      3      1.      1.     0.    1.
$
PBAR     3333   1212    50.00  104.20 416.70
$
MAT1     1212   30.+6   .333   7.-4
$
```

Exhibit 28.7 Input Bulk Data Deck

```
$
TLOAD2  2222    101             3       0.000   10.+8   1.00    .0      +LK
+LK     -1.     .0
$
$ THE "SCALING FACTOR" 10.+5 IS DETERMINED BY
$ THE VALUE OF THE BIG MASS
$
DAREA   101     1       2       -10.+5
$
$ BIG MASS  AT THE BASE
$
CMASS2  3333    10.+5   1       2
$
TSTEP   99      100     .01     1
$
$ ATTACH SPRING AND DASHPOT TO THE "FOUNDATION MASS"
$ FOR NUMERICAL STABILITY
$
CELAS2  5544    40000.  1       2
CDAMP2  331122  500.    1       2
ENDDATA
```

Exhibit 28.7 (continued) Input Bulk Data Deck

POINT-ID =   30

TIME	TYPE	T1	DISPLACEMENT VECTOR T2	T3	R1	R2	R3
0.0	G	0.0	0.0	0.0	0.0	0.0	0.0
1.000000E-02	G	0.0	3.470700E-05	0.0	0.0	0.0	0.0
2.000000E-02	G	0.0	1.218508E-04	0.0	0.0	0.0	0.0
3.000000E-02	G	0.0	2.466255E-04	0.0	0.0	0.0	0.0
4.000000E-02	G	0.0	3.350633E-04	0.0	0.0	0.0	0.0
5.000000E-02	G	0.0	3.334185E-04	0.0	0.0	0.0	0.0
6.000000E-02	G	0.0	2.411532E-04	0.0	0.0	0.0	0.0
7.000000E-02	G	0.0	1.077581E-04	0.0	0.0	0.0	0.0
8.000000E-02	G	0.0	4.272960E-06	0.0	0.0	0.0	0.0
9.000000E-02	G	0.0	-1.345478E-05	0.0	0.0	0.0	0.0

SOME LINES HAVE BEEN DELETED

TIME	TYPE	T1	T2	T3	R1	R2	R3
9.099994E-01	G	0.0	-3.635732E-05	0.0	0.0	0.0	0.0
9.199994E-01	G	0.0	8.908760E-05	0.0	0.0	0.0	0.0
9.299994E-01	G	0.0	2.014480E-04	0.0	0.0	0.0	0.0
9.399994E-01	G	0.0	2.388190E-04	0.0	0.0	0.0	0.0
9.499994E-01	G	0.0	1.794805E-04	0.0	0.0	0.0	0.0
9.599994E-01	G	0.0	5.715397E-05	0.0	0.0	0.0	0.0
9.699994E-01	G	0.0	-5.671778E-05	0.0	0.0	0.0	0.0
9.799994E-01	G	0.0	-9.740387E-05	0.0	0.0	0.0	0.0
9.899994E-01	G	0.0	-4.302861E-05	0.0	0.0	0.0	0.0
9.999993E-01	G	0.0	7.631173E-05	0.0	0.0	0.0	0.0

Exhibit 28.8   Displacement vector as a function of time

ELEMENT-ID =   10

F O R C E S   I N   B A R   E L E M E N T S   (C B A R)

	BEND-MOMENT-END-A		BEND-MOMENT-END-B		SHEAR			
TIME	PLANE 1	PLANE 2	PLANE 1	PLANE 2	PLANE 1	PLANE 2	FORCE	TORQUE
0.0	0.0	0.0	0.0	0.0	0.0	0.0	0.0	0.0
1.000000E-02	0.0	-9.883424E+01	0.0	5.229035E+00	0.0	-1.387510E+00	0.0	0.0
2.000000E-02	0.0	-2.729203E+02	0.0	-4.336755E+01	0.0	-3.060704E+00	0.0	0.0
3.000000E-02	0.0	-4.926527E+02	0.0	-1.375566E+02	0.0	-4.734615E+00	0.0	0.0
4.000000E-02	0.0	-6.445214E+02	0.0	-2.075436E+02	0.0	-5.826371E+00	0.0	0.0
5.000000E-02	0.0	-6.546987E+02	0.0	-1.954070E+02	0.0	-6.123888E+00	0.0	0.0
6.000000E-02	0.0	-4.853302E+02	0.0	-1.314972E+02	0.0	-4.717773E+00	0.0	0.0
7.000000E-02	0.0	-2.239863E+02	0.0	-5.282643E+01	0.0	-2.282131E+00	0.0	0.0
8.000000E-02	0.0	-3.997272E+01	0.0	2.381438E+01	0.0	-8.504947E-01	0.0	0.0
9.000000E-02	0.0	-1.652238E+01	0.0	4.367055E+01	0.0	-8.025725E-01	0.0	0.0
S O M E L I N E S H A V E B E E N D E L E T E D								
9.099994E-01	0.0	6.040000E+01	0.0	3.046717E+01	0.0	3.991044E-01	0.0	0.0
9.199994E-01	0.0	-1.746134E+02	0.0	-5.247714E+01	0.0	-1.628483E+00	0.0	0.0
9.299994E-01	0.0	-3.941978E+02	0.0	-1.191997E+02	0.0	-3.666641E+00	0.0	0.0
9.399994E-01	0.0	-4.513298E+02	0.0	-1.546432E+02	0.0	-3.955821E+00	0.0	0.0
9.499994E-01	0.0	-3.365117E+02	0.0	-1.184510E+02	0.0	-2.907476E+00	0.0	0.0
9.599994E-01	0.0	-1.272459E+02	0.0	-2.098071E+01	0.0	-1.416870E+00	0.0	0.0
9.699994E-01	0.0	8.474075E+01	0.0	5.543230E+01	0.0	3.907793E-01	0.0	0.0
9.799994E-01	0.0	1.730521E+02	0.0	7.226031E+01	0.0	1.343890E+00	0.0	0.0
9.899994E-01	0.0	6.240413E+01	0.0	4.362334E+01	0.0	2.504106E-01	0.0	0.0
9.999993E-01	0.0	-1.639109E+02	0.0	-3.300275E+01	0.0	-1.745443E+00	0.0	0.0

Exhibit 28.9   Forces in BAR elements as a function of time

Appendix I

List of Solution Sequences Demonstrated in the Text

Statics

SOL 24 Linear Static Analysis

$$[K]\{x\} = \{F\}$$

Dynamics

SOL 3   Normal Modes

$$([K]-\omega^2[M]) \{\phi\} = \{0\}$$

SOL 26 Direct Frequency Response

$$[K]d^2\{x\}/dt^2 + [C]d\{x\}/dt + [K]\{x\} = \{P(f)\}\exp(i\omega t)$$

SOL 27 Direct Transient Response

$$[K]d^2\{x\}/dt^2 + [C]d\{x\}/dt + [K]\{x\} = \{F(t)\}$$

SOL 31 Modal Transient Response

$$[K]d^2\{x\}/dt^2 + [K]\{x\} = \{F(t)\}$$

(damping is specified at modal level)

Appendix II

Description of Selected Bulk Data Deck Cards

This appendix presents the most commonly used Bulk Data Deck cards exactly as they appear in the standard MSC/NASTRAN documentation (User's Manual Volume I).

A number enclosed in square brackets appears next to the name of some cards. This number corresponds to the problem in which the card is used for the first time.

The author thanks The MacNeal-Schwendler Corporation for granting permission to include this information.

Bulk Data Entry: **CBAR** [7] - Single Beam Element Connection.

Description: Defines a simple beam element (BAR) of the structural model.

Format:

1	2	3	4	5	6	7	8	9	10
CBAR	EID	PID	GA	GB	X1,G0	X2	X3	blk	+x
+x	PA	PB	W1A	W2A	W3A	W1B	W2B	W3B	

Field	Contents
EID	Unique element identification number (Integer > 0).
PID	Identification number of a PBAR property entry (Default is EID unless BAROR entry has nonzero entry in field 3) (Integer > 0 or blank).*
GA, GB	Grid point identification numbers of connection points (Integer > 0; GA different than GB).
X1, X2, X3	Components of vector V, at end A, (see figure) measured at end A, parallel to the components of the displacement coordinate system for GA, to determine (with the vector from end A to end B) the orientation of the element coordinate system for the BAR element (Real, 0 or blank).
G0	Grid point identification number to optionally supply X1, X2, X3 (Integer > 0 or blank). Direction of orientation vector is GA to G0.*
PA, PB	Pin flags for bar ends A and B, respectively (up to 5 of the unique digits 1-6 anywhere in the field with no embedded blanks; Integer > 0). Used to remove connections between the grid point and the selected degrees of freedom of the bar. The degrees of freedom are defined in the element's coordinate system (see figure). The bar must have stiffness associated with the PA and PB degrees of freedom to be released by the pin flags. For example, if PA = 4 is specified, the PBAR entry must have a value for J, the torsional stiffness.
WjA, WjB	Components of the offset vectors wa and wb, respectively (see figure), in displacement coordinate systems at points GA and GB, respectively (Real or blank).

---

* See the BAROR entry for default options for fields 3 and 6 - 8.

Remarks:

1 Element identification numbers must be unique with respect to all other identification numbers.

2 For an explanation of BAR element geometry, see Section 1.3.2.

3 If there are no pin flags or offsets, the continuation may be omitted.

4 The old CBAR entry used field 9 for a flag, F, which was used to specify the nature of fields 6-8 as follows:

FIELD	6	7	8
F = 1	X1	X2	X3
F = 2	G0	Blank or 0	Blank or 0
F = blank	Provided by BAROR card		

This data item is no longer required but may be continue to be used if desired (see Remark 5). If F = 1 in field 9, a zero (0) in field 6, 7 or 8 will override entries on the BAROR entry but a blank will not.

5 For the case where field 9 is blank and not provided by the BAROR entry, if X1,G0 is integer, then G0 is used; if X1,G0 is blank or real, then X1, X2, X3 is used.

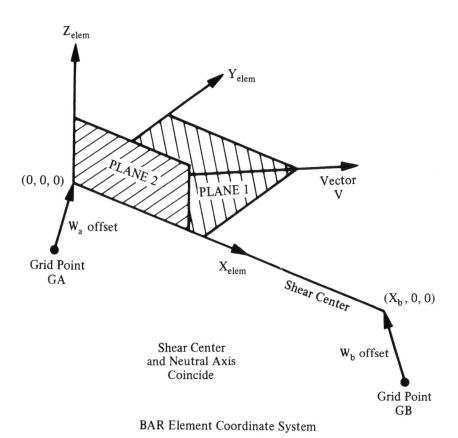

BAR Element Coordinate System

Figure A

Bulk Data Entry: **CDAMP2** [18] - Scalar Damper Property and Connection.

Description: Defines a scalar damper element of the structural model without reference to a property entry.

Format:

1	2	3	4	5	6	7	8	9	10
CDAMP2	EID	B	G1	C1	G2	C2			

Field	Content
EID	Unique element identification number (Integer > 0).
B	The value of the scalar damper (Real).
G1,G2	Geometric grid point identification number (Integer $\geq$ 0).
C1, C2	Component number (6 $\geq$ Integer $\geq$ 0).

Remarks:

1. Scalar points may be used for G1 and/or G2 in which case the corresponding C1 and/or C2 must be zero or blank. Zero or blank may be used to indicate a grounded terminal G1 or G2 with a corresponding blank or zero C1 or C2. A grounded terminal is a point whose displacement is constrained to zero.

2. Element identification numbers must be unique with respect to all other element identification numbers.

3. The single entry completely defines the elements since no material or geometric properties are required.

4. The two connection points (G1, C1) and (G2, C2) must be distinct.

5. For a discussion of the scalar elements, see Section 5.6 of the NASTRAN Theoretical Manual.

6. If this entry is used in heat transfer analysis, it generates a lumped heat capacity.

7. A scalar point specified on this entry need not be defined on an SPOINT entry.

Bulk Data Entry: **CELAS2** [10] - Scalar Spring Property and Connection.

Description: Defines a scalar spring element of the structural model without reference to a property entry.

Format:

1	2	3	4	5	6	7	8	9	10
CELAS2	EID	K	G1	C1	G2	C2	GE	S	

Field	Contents
EID	Unique element identification number (Integer > 0).
K	The value of the scalar spring (Real).
G1, G2	Geometric grid point identification number (Integer ≥ 0).
C1, C2	Component number ($6 \geq$ Integer ≥ 0).
GE	Damping coefficient (Real).
S	Stress coefficient (Real).

Remarks:

1   Scalar points may be used for G1 and/or G2 in which case the corresponding C1 and/or C2 must be zero or blank. Zero or blank may be used to indicate a grounded terminal G1 or G2 with a corresponding blank or zero C1 or C2. A grounded terminal is a point whose displacement is constrained to zero. If only scalar points and/or ground are involved, it is more efficient to use the CELAS4 entry.

2   Element identification numbers must be unique with respect to all other element identification numbers.

3   This single entry completely defines the element since no material or geometric properties are required.

4   The two connection points (G1, C1) and (G2, C2) must be distinct.

5   For a discussion of the scalar elements, see Section 5.6 of the NASTRAN Theoretical Manual.

6   A scalar point specified on this entry need not be defined on an SPOINT entry.

<u>Bulk Data Entry</u>: **CHEXA** [13] - Six-sided Solid Element with from Eight to Twenty Grid Points.

<u>Description</u>: Defines the connections of the HEXA solid element.

<u>Format</u>:

1	2	3	4	5	6	7	8	9	10
CHEXA	EID	PID	G1	G2	G3	G4	G5	G6	+x
+x	G7	G8	G9	G10	G11	G12	G13	G14	+y
+y	G15	G16	G17	G18	G19	G20			

<u>Field</u>	<u>Contents</u>
EID	Element identification number (Integer > 0).
PID	Identification number of a PSOLID property entry (Integer > 0).
G1,...,G20	Grid point identification numbers of connection points (Integer ≥ 0 or blank).

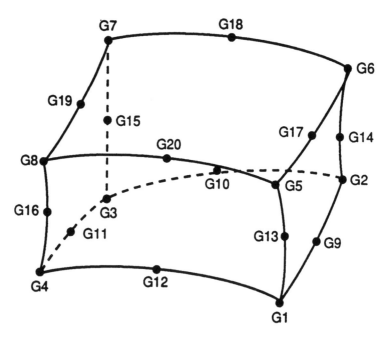

Figure B

Remarks:

1 Element identification numbers must be unique with respect to all other element identification numbers.

2 Grid points G1,...,G4 must be given in consecutive order about one quadrilateral face. G5,...,G8 must be on the opposite face with G5 opposite to G1, G6 opposite to G2, etc.

3 The edge points, G9 to G20, are optional. Any or all of them may be deleted. If the ID of any edge connection point is left blank or set to zero (as for G9 and G10 in the example), the equations of the element are adjusted to give correct results for the reduced number of connections. Corner grid points cannot be deleted. The element is an isoparametric element (with shear correction) in all cases.

4 Components of stress are output in the material coordinate system. The material coordinate system is defined on the PSOLID entry.

5 The second continuation is not required.

6 The element coordinate system for the HEXA element is defined in terms of the three face-to-face lines. Three intermediate axes R, S, and T are chosen by the following rules:

R axis: Longest line joining centroids of opposite faces.

S axis: In the plane containing longest and next longest lines joining centroids of opposite faces, and perpendicular to R.

T axis: Perpendicular to R and S.

XYZ axes: Select RST axes to make the X-axis approximately parallel to edge 1-2 and Y-axis approximately parallel to edge 1-4. In the example shown, $X = S$, $Y = R$, $Z = T$.

7 It is recommended that the edge points be located within the middle third of the edge. If the edge point is located at the quarter point, the calculated stresses will be meaningless.

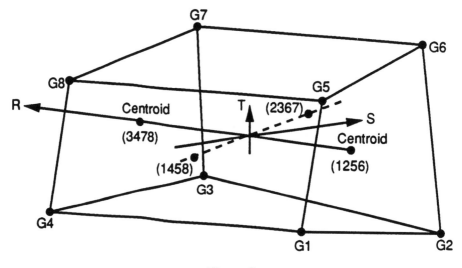

Figure C

<u>Bulk Data Entry</u>: **CMASS2** [14] - Scalar Mass Property and Connection.

<u>Description</u>: Defines a scalar mass element of the structural model without reference to a property entry.

<u>Format</u>:

1	2	3	4	5	6	7	8	9	10
CMASS2	EID	M	G1	C1	G2	C2			

Field	Contents
EID	Unique element identification number (Integer > 0).
M	The value of the scalar mass (Real).
G1, G2	Geometric grid point identification number (Integer ≥ 0).
C1, C2	Component number (6 ≥ Integer ≥ 0).

<u>Remarks</u>:

1. Scalar points may be used for G1 and/or G2 in which case the corresponding C1 and/or C2 must be zero or blank. Zero or blank may be used to indicate a grounded terminal G1 or G2 with a corresponding blank or zero C1 or C2. A grounded terminal is a point whose displacement is constrained to zero. If only scalar points and/or ground are involved, it is more efficient to use the CMASS4 entry.

2. Element identification numbers must be unique with respect to <u>all</u> other element identification numbers.

3. This single entry completely defines the element since no material or geometric properties are required.

4. The two connection points (G1, C1) and (G2, C2) must be distinct. Except in unusual circumstances, one of them will be a grounded terminal with blank entries for G and C.

5. For a discussion of the scalar elements, see Section 5.6 of the NASTRAN Theoretical Manual.

6. A scalar point specified on this entry <u>need not</u> be defined on an SPOINT entry.

<u>Bulk Data Entry</u>: **CORD1C** - Cylindrical Coordinate System Definition, Form 1.

<u>Description</u>: Defines a cylindrical coordinate system by reference to three grid points. These points must be defined in coordinate systems whose definition does not involve the coordinate system being defined. The first point is the origin, the second lies on the z-axis, and the third lies in the plane of the azimuthal origin.

<u>Format</u>:

1	2	3	4	5	6	7	8	9	10
CORD1C	CID	G1	G2	G3	CID	G1	G2	G3	

<u>Field</u>	<u>Contents</u>
CID	Coordinate system identification number (Integer > 0).
Gi	Grid point identification number (Integer > 0).

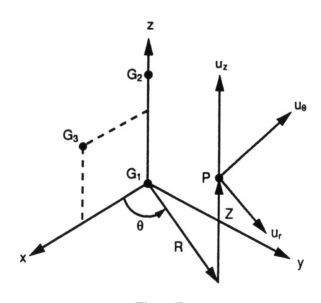

Figure D

Remarks:

1   Coordinate system identification numbers on all CORD1R, CORD1C, CORD1S, CORD2R, CORD2C and CORD2S entries must all be unique.

2   The three points G1, G2, G3 must be noncolinear.

3   The location of a grid point (P in the sketch) in this coordinate system is given by (R, θ, Z) where θ is measured in degrees.

4   The displacement coordinate directions at P are dependent on the location of P as shown by (u_r, u_θ, u_z).

5   Points on the z-axis may not have their displacement directions defined in this coordinate system since an ambiguity results.

6   One or two coordinate systems may be defined on a single entry.

<u>Bulk Data Entry</u>: **CORD1R** [24] - Rectangular Coordinate System Definition, Form 1.

<u>Description</u>: Defines a rectangular coordinate system by reference to three grid points. These points must be defined in coordinate systems whose definition does not involve the coordinate system being defined. The first point is the origin, the second lies on the z-axis, and the third lies in the x-z plane.

<u>Format</u>:

1	2	3	4	5	6	7	8	9	10
CORD1R	CID	G1	G2	G3	CID	G1	G2	G3	

<u>Field</u>	<u>Contents</u>
CID	Coordinate system identification number (Integer > 0).
Gi	Grid point identification number (Integer > 0).

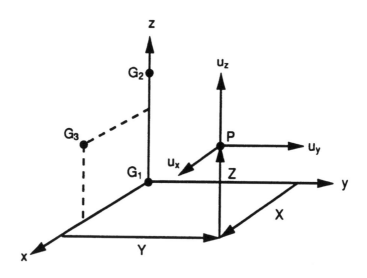

Figure E

Remarks:

1 Coordinate system identification numbers on all CORD1R, CORD1C, CORD1S, CORD2R, CORD2C and CORD2S entries must all be unique.

2 The three points G1, G2, G3 must be noncolinear.

3 The location of a grid point (P in the sketch) in this coordinate system is given by (X, Y, Z).

4 The displacement coordinate directions at P are shown by (u_x, u_y, u_z).

5 One or two coordinate systems may be defined on a single entry.

<u>Bulk Data Entry</u>: **CORD1S** - Spherical Coordinate System Definition, Form 1.

<u>Description</u>: Defines a spherical coordinate system by reference to three grid points. These points must be defined in coordinate systems whose definition does not involve the coordinate system being defined. The first point is the origin, the second lies on the z-axis, and the third lies in the plane of the azimuthal origin.

<u>Format</u>:

1	2	3	4	5	6	7	8	9	10
CORD1S	CID	G1	G2	G3	CID	G1	G2	G3	

<u>Field</u>	<u>Contents</u>
CID	Coordinate system identification number (Integer > 0).
Gi	Grid point identification number (Integer > 0).

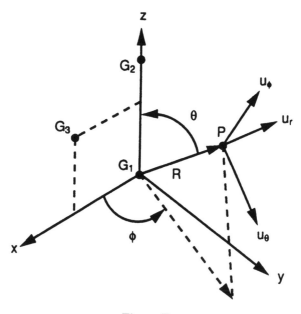

Figure F

Remarks:

1   Coordinate system identification numbers on all CORD1R, CORD1C, CORD1S, CORD2R, CORD2C and CORD2S entries must all be unique.

2   The three points G1, G2, G3 must be noncolinear.

3   The location of a grid point (P in the sketch) in this coordinate system is given by (R, θ, ϕ) where θ and ϕ are measured in degrees.

4   The displacement coordinate directions at P are dependent on the location of P as shown by (u_r, u_θ, u_ϕ).

5   Points on the polar-axis may not have their displacement directions defined in this coordinate system since an ambiguity results. In this case the basic rectangular system will be used.

6   One or two coordinate systems may be defined on a single entry.

<u>Bulk Data Entry</u>: **CORD2C** [22] - Cylindrical Coordinate System Definition, Form 2.

<u>Description</u>: Defines a cylindrical coordinate system by reference to the coordinates of three grid points. The first point defines the origin. The second point defines the direction of the z-axis. The third lies in the plane of the azimuthal origin. The reference coordinate system must be independently defined.

<u>Format</u>:

1	2	3	4	5	6	7	8	9	10
CORD2C	CID	RID	A1	A2	A3	B1	B2	B3	+x
+x	C1	C2	C3						

<u>Field</u>	<u>Contents</u>
CID	Coordinate system identification number (Integer > 0).
RID	Reference to a coordinate system which is defined independently of new coordinate system (Integer $\geq$ 0 or blank).
Ai, Bi,Ci	Coordinates of three points in coordinate system defined in field 3 (Real).

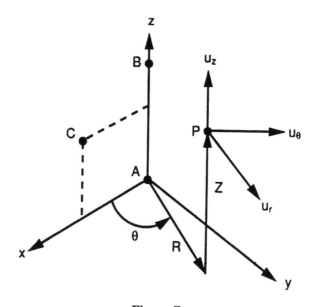

Figure G

Remarks:

1   Continuation entry must be present.

2   The three points (A1, A2, A3), (B1, B2, B3), (C1, C2, C3) must be unique and noncolinear. Noncolinearity is checked by the geometry processor.

3   Coordinate system identification numbers on all CORD1R, CORD1C, CORD1S, CORD2R, CORD2C and CORD2S entries must all be unique.

4   An RID of zero references the basic coordinate system.

5   The location of a grid point (P in the sketch) in this coordinate system is given by (R, θ, Z) where θ is measured in degrees.

6   The displacement coordinate directions at P are dependent on the location of P as shown by (u_r, u_θ, u_z).

7   Points on the z-axis may not have their displacement directions defined in this coordinate system since an ambiguity results. In this case, the basic rectangular system will be used.

<u>Bulk Data Entry</u>: **CORD2R** - Rectangular Coordinate System Definition, Form 2.

<u>Description</u>: Defines a rectangular coordinate system by reference to the coordinates of three grid points. The first point defines the origin. The second point defines the direction of the z-axis. The third point defines a vector which, with the z-axis, defines the x-z plane. The reference coordinate system must be independently defined.

<u>Format</u>:

1	2	3	4	5	6	7	8	9	10
CORD2R	CID	RID	A1	A2	A3	B1	B2	B3	+x
+x	C1	C2	C3						

<u>Field</u>	<u>Contents</u>
CID	Coordinate system identification number (Integer > 0).
RID	Reference to a coordinate system which is defined independently of new coordinate system (Integer ≥ 0 or blank).
Ai, Bi,Ci	Coordinates of three points in coordinate system defined in field 3 (Real).

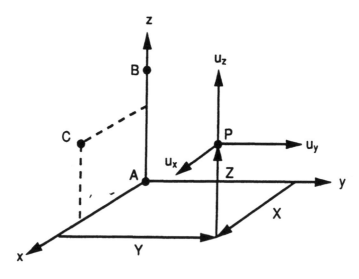

Figure H

Remarks:

1   Continuation entry <u>must</u> be present.

2   The three points (A1, A2, A3), (B1, B2, B3), (C1, C2, C3) must be unique and noncolinear. Noncolinearity is checked by the geometry processor.

3   Coordinate system identification numbers on all CORD1R, CORD1C, CORD1S, CORD2R, CORD2C and CORD2S entries must all be unique.

4   An RID of zero references the basic coordinate system.

5   The location of a grid point (P in the sketch) in this coordinate system is given by (X, Y, Z).

6   The displacement coordinate directions at P are shown by (u_x, u_y, u_z).

<u>Bulk Data Entry</u>: **CORD2S** - Spherical Coordinate System Definition, Form 2.

<u>Description</u>: Defines a rectangular coordinate system by reference to the coordinates of three grid points. The first point defines the origin. The second point defines the direction of the z-axis. The third lies in the plane of the azimuthal origin. The reference coordinate system must be independently defined.

<u>Format</u>:

1	2	3	4	5	6	7	8	9	10
CORD2S	CID	RID	A1	A2	A3	B1	B2	B3	+x
+x	C1	C2	C3						

<u>Field</u>	<u>Contents</u>
CID	Coordinate system identification number (Integer > 0).
RID	Reference to a coordinate system which is defined independently of new coordinate system (Integer $\geq$ 0 or blank).
Ai, Bi,Ci	Coordinates of three points in coordinate system defined in field 3 (Real).

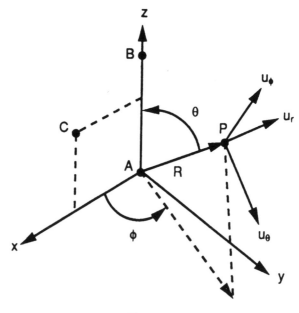

Figure I

Remarks:

1   Continuation entry <u>must</u> be present.

2   The three points (A1, A2, A3), (B1, B2, B3), (C1, C2, C3) must be unique and noncolinear. Noncolinearity is checked by the geometry processor.

3   Coordinate system identification numbers on all CORD1R, CORD1C, CORD1S, CORD2R, CORD2C and CORD2S entries must all be unique.

4   An RID of zero references the basic coordinate system.

5   The location of a grid point (P in the sketch) in this coordinate system is given by (R, θ, ϕ) where θ and ϕ are measured in degrees.

6   The displacement coordinate directions at P are shown by (u_r, u_θ, u_ϕ).

7   Points on the polar axis may not have their displacement directions defined in this coordinate system since an ambiguity results.

<u>Bulk Data Entry</u>: **CPENTA** - Five-sided Solid Element with from 6 to 15 grid points.

<u>Description</u>: Defines the connections of the CPENTA element.

<u>Format</u>:

1	2	3	4	5	6	7	8	9	10
CPENTA	EID	PID	G1	G2	G3	G4	G5	G6	+x
+x	G7	G8	G9	G10	G11	G12	G13	G14	+y
+y	G15								

Field	Contents
EID	Element identification number (Integer > 0).
PID	Identification number of a PSOLID property entry (Integer > 0).
G1-G15	Identification numbers of connected grid points (Integer ≥ 0 or blank).

Figure J

Remarks:

1   Element ID numbers must be unique with respect to all other element ID numbers.

2   The topology of the diagram must be preserved, i.e., G1, G2, G3 define a triangular face, G1, G10 and G4 are on the same edge, etc.

3   The edge of grid points, G7 to G15, are optional. Any or all of them may be deleted. In the example shown, G10, G11 and G12 have been deleted. The continuations are not required if all edge grid points are deleted.

4   Components of stress are output in the material coordinate system.

5   The element coordinate system is defined as follows:

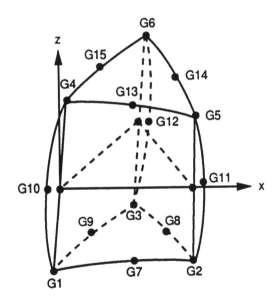

Figure K

where the x-axis joins the midpoints of straight lines joining points G1-G4 and G2-G5, and the z-axis is normal to a plane passing through the midpoints of straight lines joining G1-G4, G2-G5 and G3-G6.

6   It is recommended that the edge points be located within the middle third of the edge. If the edge point is located at the quarter point, the calculated stresses will be meaningless.

<u>Bulk Data Entry</u>: **CQUAD4** [10] - Quadrilateral Element Connection.

<u>Description</u>: Defines a quadrilateral plate element (QUAD4) of the structural model. This is an isoparametric membrane-bending element.

<u>Format</u>:

1	2	3	4	5	6	7	8	9	10
CQUAD4	EID	PID	G1	G2	G3	G4	THETA	ZOFFS	+x
+x	blk	blk	T1	T2	T3	T4			

<u>Field</u>	<u>Contents</u>
EID	Element identification number (Unique integer > 0).
PID	Identification number of a PSHELL or PCOMP property entry (Integer > 0 or blank, default is EID).
Gi	Grid point identification numbers of connection points (Integers > 0, all unique).
THETA	Material property orientation specification (Real or blank; or 0 ≤ Integer < 1,000,000). If Real or blank, specifies the material property orientation angle in degrees. If integer, the orientation of the material x-axis is along the projection onto the plane of the element of the x-axis of the coordinate system specified by the integer value. The sketch gives the sign convention for θ.
ZOFFS	Offset from the surface of grid points to the element reference plane (see Remark 6) (Real).
Ti	Membrane thickness of element at grid points G1 through G4. (Real ≥ 0. or blank, not all zero. See Remark 4 for default.)

<u>Remarks</u>:

1 Element identification numbers must be unique with respect to all other element identification numbers.

2 Grid points G1 through G4 must be ordered consecutively around the perimeter of the element.

3 All the interior angles must be less than 180 degrees.

4 The continuation is optional. If it is not supplied, then T1 through T4 will be set equal to the value of T on the PSHELL data entry.

5 Stresses are output in the element coordinate system.

6   Elements may be offset from the grid point surface by means of ZOFFS. Other data, such as material matrices and stress fiber locations are given relative to the reference plane. Positive offset implies that the element reference plane lies above the grid points in the sketch.

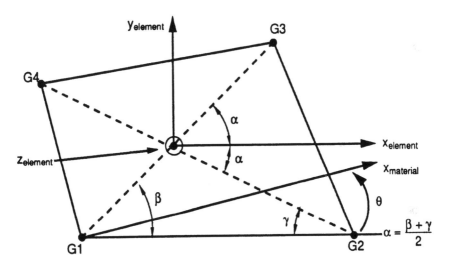

Figure L

Bulk Data Entry: **CQUAD8** [12] - Quadrilateral Element Connection.

Description: Defines a curved quadrilateral shell element (QUAD8) with eight grid points.

Format:

1	2	3	4	5	6	7	8	9	10
CQUAD8	EID	PID	G1	G2	G3	G4	G5	G6	+x
+x	G7	G8	T1	T2	T3	T4	THETA	ZOFFS	

Field	Contents
EID	Element identification number (Integer > 0).
PID	Identification number of a PSHELL or PCOMP property entry (Integer > 0).
G1,...,G4	Identification numbers of connected corner grid points (Unique integers > 0). Required data for all four grid points.
G5,...,G8	Identification numbers of connected edge grid points (Integer ≥ 0 or blank). Optional data for any or all four grid points.
Ti	Membrane thickness of element at corner grid points. (Real ≥ 0. or blank, not all zero.) See Remark 4 for default.
THETA	Material property orientation specification (Real or blank; or $0 \leq$ Integer $< 1,000,000$). If Real or blank, specifies the material property orientation angle in degrees. If integer, the orientation of the material x-axis is along the projection onto the plane of the element of the x-axis of the coordinate system specified by the integer value. The sketch gives the sign convention for θ.
ZOFFS	Offset from the surface of grid points to the element reference plane (see Remark 6) (Real).

Remarks:

1   Element identification numbers must be unique with respect to all element IDs of any kind.

2   Grid points G1 to G8 must be numbered as shown.

3   The material property orientation angle, θ, is defined locally at each interior integration point as the angle made by the x-axis of the material coordinate system with a line, $\eta =$ const. If the shape of the element is a parallelogram and

if the edge points are located at midpoints of the sides, the lines $\eta = $ const. are parallel to side 1-2.

4   T1, T2, T3 and T4 are optional. If not supplied they will be set equal to the value of T on the PSHELL entry.

5   It is recommended that all the edge points be located within the middle third of the edge. If the edge point is located at the quarter point, the program may fail with a divide by zero or the calculated stresses will be meaningless.

6   Elements may be offset from the grid point surface by means of ZOFFS. Other data, such as material matrices and stress fiber locations are given relative to the reference plane. Positive offset implies that the element reference plane lies above the grid points in the sketch.

7   If all midside nodes are deleted, the element is excessively stiff. A user warning message is printed. A QUAD4 element is recommended, instead.

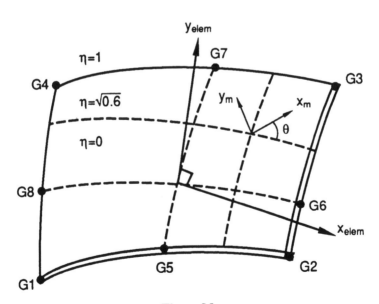

Figure M

Bulk Data Entry: **CROD** [1] - Rod Element Connection.

Description: Defines a tension-compression-torsion element (ROD) of the structural model.

Format:

1	2	3	4	5	6	7	8	9	10
CROD	EID	PID	G1	G2					

Field	Contents
EID	Element identification number (Integer > 0).
PID	Identification number of a PROD property entry (Default is EID) (Integer > 0).
G1, G2	Grid point identification numbers of connection points (Integer > 0).

Remarks:

1   Element identification numbers must be unique with respect to all other element identification numbers.

2   See CONROD for alternative method of rod definition.

3   Only one ROD element may be defined on a single entry.

<u>Bulk Data Entry</u>: **CTETRA** - Four-sided Solid Element with from 4 to 10 Grid Points.

<u>Description</u>: Defines the connections of the TETRA element.

<u>Format</u>:

1	2	3	4	5	6	7	8	9	10
CTETRA	EID	PID	G1	G2	G3	G4	G5	G6	+x
+x	G7	G8	G9	G10					

<u>Field</u>	<u>Contents</u>
EID	Element identification number (Integer > 0).
PID	Identification number of a PSOLID property entry (Integer > 0).
G1-G10	Identification numbers of connected grid points (Integer ≥ 0 or blank).

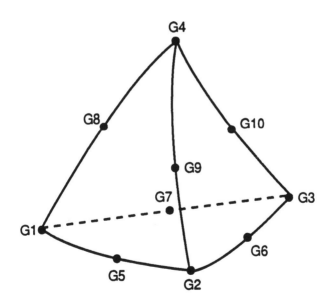

Figure N

Remarks:

1   Element ID numbers must be unique with respect to all other element ID numbers.

2   The topology of the diagram must be preserved, i.e. G1, G2, G3 define a triangular face; G1, G8 and G4 are on the same edge, etc.

3   The edge points, G5 to G10, are optional. Any or all of them may be deleted. If the ID of any edge connection point is left blank or set to zero, the equations of the element are adjusted to give correct results for the reduced number of connections. Corner grid points cannot be deleted. The element is an isoparametric element in all cases.

4   Components of stresses are output in the material coordinate system.

5   The element coordinate system is defined as follows:

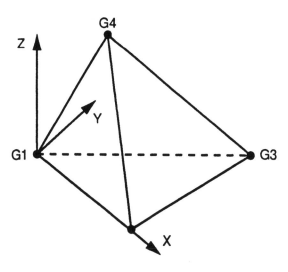

Figure O

The origin is located at G1 and the x-axis lies on the G1-G2 edge. The y-axis lies in the G1-G2-G3 plane and is perpendicular to the x-axis. The positive y-axis lies on the same side of the G1-G2 edge as node G3. The z-axis is orthogonal to the x and y axes.

6   It is recommended that the edge points be located within the middle third of the edge. If the edge point is located at the quarter point, the calculated stresses will be meaningless.

7   The constant strain element available prior to Version 64 is replaced with an improved isoparametric version. Models that used the older form of this entry may be converted by changing field 3 from a material ID to a property ID for a PSOLID entry.

Bulk Data Entry: **CTRIA3** [11] - Triangular Element Connection.

Description: Defines a triangular plate element (TRIA3) of the structural model. This is an isoparametric membrane-bending element.

Format:

1	2	3	4	5	6	7	8	9	10
CTRIA3	EID	PID	G1	G2	G3	THETA	ZOFFS	blk	+x
+x	blk	blk	T1	T2	T3				

Field	Contents
EID	Element identification number(Unique integer > 0).
PID	Identification number of a PSHELL or PCOMP property entry (Integer > 0 or blank, default is EID).
Gi	Grid point identification numbers of connection points (Integer > 0, all unique).
THETA	Material property orientation specification (Real or blank; or $0 \leq$ Integer $< 1,000,000$). If Real or blank, specifies the material property orientation angle in degrees. If integer, the orientation of the material x-axis is along the projection onto the plane of the element of the x-axis of the coordinate system specified by the integer value. The sketch gives the sign convention for THETA.
ZOFFS	Offset from the surface of grid points to the element reference plane (see Remark 3) (Real).
Ti	Membrane thickness of element at grid points G1, G2, and G3 (Real $\geq$ .0 or blank, not all zero. See Remark 2 for default.)

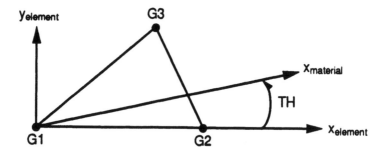

Figure P

<u>Remarks</u>:

1   Element identification numbers must be unique with respect to all other element identification numbers.

2   The continuation is optional. If it is not supplied, then T1 through T3 will be set equal to the value of T on the PSHELL data entry.

3   Elements may be offset from the grid point surface by means of ZOFFS. Other data, such as material matrices and stress fiber locations are given relative to the reference plane. Positive offset implies that the element reference plane lies above the grid points in the sketch.

<u>Bulk Data Entry</u>: **CTRIA6** - Triangular Element Connection.

<u>Description</u>: Defines a curved triangular shell element (TRIA6) with six grid points.

<u>Format</u>:

1	2	3	4	5	6	7	8	9	10
CTRIA6	EID	PID	G1	G2	G3	G4	G5	G6	+x
+x	THETA	ZOFFS	T1	T2	T3				

Field	Contents
EID	Element identification number (Integer > 0).
PID	Identification number of PSHELL or PCOMP property entry (Integer > 0).
G1 to G3	Identification number of connected corner grid points (Unique integers > 0).
G4 to G6	Identification number of connected edge grid points (Integer $\geq$ 0 or blank). Optional data for any or all three points.
THETA	Material property orientation specification (Real or blank; or $0 \leq$ Integer $< 1,000,000$). If Real or blank, specifies the material property orientation angle in degrees. If integer, the orientation of the material x-axis is along the projection onto the plane of the element of the x-axis of the coordinate system specified by the integer value. The sketch gives the sign convention for θ.
ZOFFS	Offset from the surface of grid points to the element reference plane (see Remark 6) (Real).
Ti	Membrane thickness of element at corner grid points. (Real $\geq$ 0., or blank, not all zero. See Remark 4 for default.)

<u>Remarks</u>:

1 Element identification numbers must be unique with respect to all other element IDs of any kind.

2 Grid points G1 to G6 must be numbered as shown.

3 The material property orientation angle, θ, is defined locally at each interior integration point as the angle made by the x-axis of the material coordinate system with a line, $\eta = $ const. If the sides are straight and if the edge grid points are located at the midpoints of the sides, the lines $\eta = $ const. are parallel to side 1-2.

4   T1, T2 and T3 are optional. If not supplied they will be set equal to the value of T on the PSHELL entry.

5   It is recommended that the edge points be located within the middle third of the edge. If the edge point is located at the quarter point, the program may fail with a divide by zero and the calculated stresses will be meaningless.

6   Elements may be offset from the grid point surface by means of ZOFFS. Other data, such as material matrices and stress fiber locations are given relative to the reference plane. Positive offset implies that the element reference plane lies above the grid points in the sketch.

7   If all midside nodes are deleted, the element is excessively stiff. A user warning message is printed. A TRIA3 element is recommended instead.

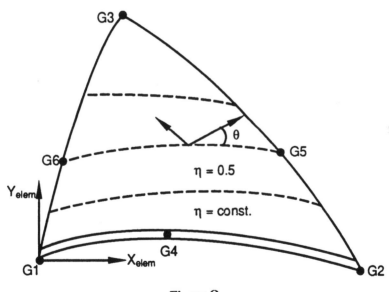

Figure Q

Bulk Data Entry: **DAREA** [18] - Dynamic Load Scale Factor.

Description: This entry is used in conjunction with the RLOAD1, RLOAD2, TLOAD1, and TLOAD2 data entries and defines the point where the dynamic load is to be applied with the scale (area) factor A.

Format:

1	2	3	4	5	6	7	8	9	10
DAREA	SID	P	C	A	P	C	A	blk	

Field	Contents
SID	Identification number of DAREA set (Integer > 0).
P	Grid, extra point or scalar point identification number (Integer > 0).
C	Component number (1-6 for grid point; blank or 0 for extra point or scalar point).
A	Scale (area) factor A for the designated coordinate (Real).

Remarks:

1 One or two scale factors may be defined on a single entry.

2 Refer to RLOAD1, RLOAD2, TLOAD1 or TLOAD2 entries for the formulas which define the scale factor A.

3 Component numbers refer to global coordinates.

4 DAREA entries may be replaced or used with LSEQ Bulk Data entries.

5 If DAREA is called by a GUST entry, P (Grid Point) must be entered as any legitimate point. However, it is only used if selected through a DLOAD request. WG from the GUST entry is used instead of A when requested via a GUST request.

Bulk Data Entry: **DEFORM** [4] - Element Deformation.

Description: Defines enforced axial deformation for one-dimensional elements for use in statics problems.

Format:

1	2	3	4	5	6	7	8	9	10
DEFORM	SID	EID	D	EID	D	EID	D	blk	

Field	Contents
SID	Deformation set identification number (Integer > 0).
EID	Element number (Integer > 0).
D	Deformation (+ = elongation) (Real).

Remarks:

1. The referenced element must be one-dimensional (i.e. a ROD (including CONROD), TUBE or BAR).

2. Deformations sets must be selected in the Case Control Section (DEFORM = SID) to be used by MSC/NASTRAN.

3. From one to three enforced element deformations may be defined on a single entry.

Bulk Data Entry: **DELAY** [20] - Dynamic Load Time Delay.

Description: This entry is used in conjunction with the RLOAD1, RLOAD2, TLOAD1 and TLOAD2 data entries and defines the time delay term τ in the equations of the loading function.

Format:

1	2	3	4	5	6	7	8	9	10
DELAY	SID	P	C	T	P	C	T		

Field	Contents
SID	Identification number of DELAY set (Integer > 0).
P	Grid, extra point or scalar point identification number (Integer > 0).
C	Component number (1-6 for grid point, blank or 0 for extra point or scalar point).
T	Time delay τ for designated coordinate (Real).

Remarks:

1   One or two dynamic load time delays may be defined on a single entry.

2   Refer to RLOAD1, RLOAD2, TLOAD1 or TLOAD2 entries for the formulas which define the manner in which the time delay, τ, is used.

3   A DAREA entry should be defined for the same grid point and component.

4   In superelement analysis DELAY entries may only be applied to loads on points in the residual structure.

Bulk Data Entry: **DLOAD** [20] - Dynamic Load Combination (Superposition).

Description: Defines a dynamic loading condition for frequency response or transient response problems as a linear combination of load sets defined via RLOAD1 or RLOAD2 entries (for frequency response) or TLOAD1 or TLOAD2 entries (for transient response).

Format:

1	2	3	4	5	6	7	8	9	10
DLOAD	SID	S	S1	L1	S2	L2	S3	L3	+x
+x	S4	L4	etc.						

Field	Contents
SID	Load set identification number (Integer > 0).
S	Scale factor (Real).
Si	Scale factors (Real).
Li	Load set identification numbers defined via entry types enumerated above (Integer > 0).

Remarks:

1. The load vector being defined by this entry is given by

$$\{P\} = S\Sigma_i Si\{Pi\}$$

2. The Li must be unique.

3. SID must be unique for all Li.

4. Nonlinear transient loads may not be included; they are selected separately in the Case Control Section.

5. Linear load sets must be selected in the Case Control Section (DLOAD = SID) to be used by MSC/NASTRAN.

6. A DLOAD entry may not reference a set identification number defined by another DLOAD entry.

7. TLOAD1 and TLOAD2 loads may be combined only through the use of the DLOAD entry.

8. RLOAD1 and RLOAD2 loads may be combined only through the use of the DLOAD entry.

9. SID must be unique for all TLOAD1, TLOAD2, RLOAD1 and RLOAD2 entries.

<u>Bulk Data Entry</u>: **EIGR** [14] - Real Eigenvalue Extraction Data.

<u>Description</u>: Defines data needed to perform real eigenvalue analysis.

<u>Format</u>:

1	2	3	4	5	6	7	8	9	10
EIGR	SID	METHOD	F1	F2	NE	ND	blk	blk	+x
+x	NORM	G	C						

Field	Contents
SID	Set identification number (Unique integer > 0).
METHOD	Method of eigenvalue extraction, one of the BCD values, "INV", "SINV", "GIV", "MGIV", etc.

	INV	Inverse power method
	SINV	Inverse power method with enhancements
	GIV	Givens' method of tridiagonalization
	MGIV	Modified Givens' method
	HOU	Householder's method of tridiagonalization
	MHOU	Modified Householder's Method
	AGIV	Automatic selection of METHOD=GIV or MGIV. (See Remark 15)
	AHOU	Automatic selection of METHOD=HOU or MHOU. (See Remark 15)

When METHOD = INV or SINV:

F1, F2	Frequency range of interest (Real ≥ 0.0). Both must be input.
NE	Estimate of number of roots in the range (Required for METHOD = "INV"). (Integer > 0). Not used by SINV METHOD.
ND	Desired number of roots. (Default is 3*NE, INV only) (Integer > 0). If blank, all roots between F1 and F2 are searched for (SINV only). Limit is 600 roots.

When METHOD = GIV, MGIV, HOU or MHOU:

F1, F2	Frequency range of interest (Real $\geq$ 0.0; F1 < F2). If ND is not blank, F1 and F2 are ignored. If ND is blank, eigenvectors are found whose natural frequencies lie in the range between F1 and F2.
NE	Not used.
ND	Desired number of eigenvectors. (Integer $\geq$ 0). If ND is zero, the number of eigenvectors is determined from F1 and F2. (Default =0). If all three are blank, then ND is automatically set to one more than the number of degrees of freedom listed on SUPORT entries.

NORM   Method for normalizing eigenvectors, one of the BCD values, "MASS", "MAX" or "POINT".

MASS   Normalize to unit value of the generalized mass (Default)

MAX   Normalize to unit value of the largest component in the analysis set

POINT   Normalize to unit value of the component defined in fields 3 and 4 (defaults of "MAX" if defined component is zero)

G   Grid or scalar point identification number (Required only if NORM="POINT") (Integer > 0).

C   Component number (One of integers 1-6) (Required only if NORM="POINT" and G is a geometric grid point).

Remarks:

1   See Section 4.2.4 of The Handbook for Dynamic Analysis for a discussion of method of selection.

2   Real eigenvalue extraction data sets must be selected in the Case Control Section (METHOD=SID) to be used by MSC/NASTRAN.

3   The units of F1 and F2 are cycles per unit of time.

4   The continuation is not required. Mass normalization is then used.

5   If METHOD="GIV" or "MGIV", all eigenvalues are found.

6   If METHOD="GIV" or "MGIV", the mass matrix for the analysis must be positive definite. The auto-omit feature removes massless degrees of freedom. Singularities or near-singularities of the mechanism type in the mass matrix will produce poor numerical stability for the GIV method. The MGIV method should be used for this condition.

7   MGIV is a modified form of the GIV method that allows a non-positive definite mass matrix for the analysis set (i.e., massless degrees of freedom may exist in the analysis set). The MGIV method should give improved accuracy for the lowest frequency solutions.

8   If NORM=MAX, components that are not in the analysis set may have values larger than unity.

9   If NORM=POINT, the selected component should be in the analysis set. The program uses MAX when it is not in the analysis set.

10   The desired number of roots (ND) includes all roots previously found, such as rigid body modes determined with the use of the SUPORT entry, or the number of roots found on the previous run when restarting and APPENDing the eigenvector file.

11   The SINV method is an enhanced version of the INV method. It uses Sturm sequence number techniques to make it more likely that all roots in the range have been found. It is generally more reliable and more efficient than the INV method.

12   For the INV and SINV method, convergence is achieved at 1.0E-6. For the other methods convergence is not tested.

13   For the SINV method only, if F2 is blank, the first shift will be made at F1, and only one eigensolution above F1 will be calculated. If there are no modes below F1, it is likely that the first mode will be calculated. If there are modes below F1 (including rigid body modes defined by SUPORT entries), a mode higher than the first mode above F1 may be calculated.

14   The HOU and MHOU methods are new methods introduced in Version 64. They correspond to the GIV and MGIV methods, so that all remarks to these older methods also pertain to the HOU and MHOU methods, respectively. The new methods generally are faster than the old methods, and will someday replace them.

15   If METHOD = AGIV or AHOU, the program automatically determines the need for a modified method (MGIV or MHOU) and make the proper selection.

Bulk Data Entry: **FORCE** [1] - Static Load.

Description: Defines a static load at a grid point by specifying a vector.

Format:

1	2	3	4	5	6	7	8	9	10
FORCE	SID	G	F	CID	N1	N2	N3	AXI	

Field	Contents
SID	Load set identification number (Integer > 0).
G	Grid point identification number (Integer > 0).
CID	Coordinate system identification number (Integer $\geq$ 0) (Default = 0).
F	Scale Factor (Real).
Ni	Components of Vector measured in coordinate system defined by CID (Real; must have at least one nonzero component).
AXI	Indicates an axisymmetric loading (BDC: "AXI" or blank).

Remarks:

1. The static load applied to grid point G is given by

$$f = F N$$

where N is the vector defined in fields 6, 7 and 8.

2. Load sets must be selected in the Case Control Section (LOAD = SID) to be used by MSC/NASTRAN.

3. A CID of zero references the basic coordinate system.

4. For the axisymmetric loading, MSC/NASTRAN calculates

$$F_{total} = 2\pi R f$$

where R is the radius of the grid point from the axis of symmetry. This is a convenience feature for use with the TRIAX6 element.

Bulk Data Entry: **FREQ** - Frequency List.

Description: Defines a set of frequencies to be used in the solution of frequency response problems.

Format:

1	2	3	4	5	6	7	8	9	10
FREQ	SID	F	F	F	F	F	F	F	+x
+x	F	F	etc.						

Field	Contents
SID	Frequency set identification number (Integer > 0).
F	Frequency value (Real > 0.0)

Remarks:

1. The units for the frequencies are cycles per unit of time.

2. Frequency sets must be selected in the Case Control Section (FREQ = SID) to be used by MSC/NASTRAN.

3. All FREQ, FREQ1 and FREQ2 entries with the same frequency set identification numbers will be used. Duplicate frequencies will be ignored. f_n and f_{n-1} are considered duplicated if $|f_n - f_{n-1}| < 10.\text{E-5} \, |f_{max} - f_{min}|$.

Bulk Data Entry: **FREQ1** [21] - Frequency List, Alternate Form 1.

Description: Defines a set of frequencies to be used in the solution of frequency response problems by specification of a starting frequency, frequency increment, and number of increments desired.

Format:

1	2	3	4	5	6	7	8	9	10
FREQ1	SID	F1	DF	NDF					

Field	Contents
SID	Frequency set identification number (Integer > 0).
F1	First frequency in set (Real > 0.0)
DF	Frequency increment (Real > 0.0)
NDF	Number of frequency increments (Integer > 0).

Remarks:

1. The units for the frequency F1 and the frequency increment DF are cycles per unit time.

2. The frequencies defined by this entry are given by

    $$f_i = F1 + (i-1)DF, \quad i = 1, NDF + 1$$

3. Frequency sets must be selected in the Case Control Section (FREQ = SID) to be used by MSC/NASTRAN.

4. All FREQ, FREQ1 and FREQ2 entries with the same frequency set identification numbers will be used. Duplicate frequencies will be ignored. f_n and f_{n-1} are considered duplicated if $|f_n - f_{n-1}| < 10.E-5 |f_{max} - f_{min}|$.

Bulk Data Entry: **GRAV** [7] - Gravity Vector.

Description: Used to define gravity vectors for use in determining gravity loading for the structural model.

Format:

1	2	3	4	5	6	7	8	9	10
GRAV	SID	CID	G	N1	N2	N3			

Field	Contents
SID	Set identification number (Integer > 0).
CID	Coordinate system identification number (Integer > 0).
G	Gravity vector scale factor.
Ni	Gravity vector components (Real; at least one nonzero component).

Remarks:

1. The gravity vector is defined by g = G(N1, N2, N3). The direction of g is the direction of free fall. N1, N2, N3 are in coordinate system CID.

2. A CID of zero references the basic coordinate system.

3. Gravity loads may be combined with "simple loads" (e.g., FORCE, MOMENT) only by specification on a LOAD entry. That is, the SID on a GRAV entry may not be the same as that on a simple entry.

4. Load sets must be selected in the Case Control Section (LOAD = SID) to be used by MSC/NASTRAN.

5. At most nine GRAV entries can be selected in a given run either by Case Control or the LOAD Bulk Data entry.

6. In cyclic symmetry solution sequences, the T3 axis of the coordinate system referenced in field 3 must be parallel to the axis of symmetry. In the DIH type of cyclic symmetry the T1 axis must, in addition, be parallel to Side 1 of segment 1R of the model.

7. For image superelements, the coordinate system must be rotated if the image is rotated relative to its primary superelement.

Bulk Data Entry: **GRDSET** - Grid Point Default.

Description: Defines default options for fields 3, 7 and 8 of all GRID entries.

Format:

1	2	3	4	5	6	7	8	9	10
GRDSET	blk	CP	blk	blk	blk	CD	PS	blk	

Field	Contents
CP	Identification number of coordinate system in which the location of the grid point is defined (Integer ≥ 0).
CD	Identification number of coordinate system in which the displacements are measured at grid point (Integer ≥ 0).
PS	Permanent single-point constraints associated with grid point (any of the digits 1-6 with no embedded blanks) (Integer ≥ 0).

Remarks:

1. The contents of field 3, 7 or 8 of this entry are assumed for the corresponding fields of any GRID entry whose field 3, 7 and 8 are blank. If any of these fields on the GRID entry are blank, the default option defined by this entry occurs in that field. If no permanent single-point constraints are desired or one of the coordinate system is basic, the default may be overridden on the GRID entry by making one of the fields 3, 7 or 8 zero (rather than blank). Only one GRDSET entry may appear in the user's Bulk Data section.

2. The primary purpose of this entry is to minimize the burden of preparing data for problems with a large amount of repetition (e.g., two-dimensional pinned-joint problems).

3. At least one of the entries CP, CD or PS must be nonzero.

Bulk Data Entry: **GRID** [1] - Grid Point.

Description: Defines the location of a geometric grid point of the structural model, the directions of its displacement, and its permanent single-point constraint.

Format:

1	2	3	4	5	6	7	8	9	10
GRID	ID	CP	X1	X2	X3	CD	PS	SEID	

Field	Contents

ID   Grid point identification number
            $(1,000,000 > \text{Integer} > 0)$.

CP   Identification number of coordinate system in which the location of the grid point is defined (Integer ≥ 0).

Xi   Location of the grid point in coordinate system CP (Real).

CD   Identification number of coordinate system in which the displacements, degrees of freedom, constraints, and solution vectors are defined at grid point (Integer ≥ 0 or blank)[*]

PS   Permanent single-point constraints associated with grid point (any of the digits 1-6 with no imbedded blanks) (Integer ≥ 0 or blank)[*]

SEID   Superelement identification number (Integer ≥ 0 or blank).

Remarks:

1   All grid point identification numbers must be unique with respect to all other structural, scalar and fluid points.

2   The meaning of X1, X2 and X3 depend on the type of coordinate system, CP, as follows: (see CORDi entry description).

---

[*] See the GRDSET entry for default options for fields 3, 7 and 8.

Type	X1	X2	X3
Rectangular	X	Y	Z
Cylindrical	R	θ (degrees)	Z
Spherical	R	θ (degrees)	ϕ (degrees)

3   The collection of all CD coordinate systems defined on all GRID entries is called the Global Coordinate System. All degrees of freedom, constraints, and solution vectors are expressed in the Global Coordinate System.

4   The SEID entry can be override by the use of the SESET Bulk Data entry.

Bulk Data Entry: **LOAD** [6] - Static Load Combination (Superposition).

Description: Defines a static load as a linear combination of load sets defined via FORCE, MOMENT, FORCE1, MOMENT1, FORCE2, MOMENT2, PLOAD, PLOAD1, PLOAD2, PLOAD3, PLOAD4, PLOADX, SLOAD, RFORCE, and GRAV entries.

Format:

1	2	3	4	5	6	7	8	9	10
LOAD	SID	S	S1	L1	S2	L2	S3	L3	+x
+x	S4	L4	etc.						

Field	Contents
SID	Load set identification number (Integer > 0).
S	Scale factor (Real).
Si	Scale factors (Real).
Li	Load set identification numbers defined via entry types enumerated above (Integer > 0).

Remarks:

1. The load vector defined is given by

$$\{P\} = S\Sigma S_i \{P_{Li}\}$$

2. The Li must be unique. The remainder of the physical entry containing the last entry must be blank.

3. This entry must be used if gravity loads (GRAV) are to be used with any of the other types.

4. Load sets must be selected in the Case Control Section (LOAD = SID) if they are to be applied to the structural model.

5. A LOAD entry may not reference a set identification number defined by another LOAD entry.

6. There may be at most 300 (Si, Li) pairs.

<u>Bulk Data Entry</u>: **LSEQ** [19} - Static Load Set Definition.

<u>Description</u>: Defines a sequence of static load sets to be applied to the structural model. The load sets may be referenced by dynamic load entries.

<u>Format</u>:

1	2	3	4	5	6	7	8	9	10
LSEQ	SID	DAREA	LID	TID					

Field	Contents
SID	Set identification of the set of LSEQ entries (Integer > 0).
DAREA	The DAREA set identification assigned to this static load vector (Integer > 0).
LID	Load set identification number of a set of static load entries. (Any entry that may be referenced by the LOAD Case Control command) (Integer > 0 or blank).
TID	Set identification of a thermal load set (Any entry that may be referenced by the TEMP(LOAD) Case Control command) (Integer > 0 or blank).

<u>Remarks</u>:

1. The above entries will not be used unless selected in the Case Control Section with a LOADSET entry.

2. This entry is available only in superelements and Solution Sequences 26, 27, 30 and 31.

3. The number of static load vectors created for each superelement is the number of unique DAREA IDs on all LSEQ entries in the bulk data.

4. The DAREA identification assigned to the static load vectors may be referenced by CLOAD, RLOAD1, RLOAD2, TLOAD1 and TLOAD2 entries.

5. Element data recovery for thermal loads is not currently implemented in dynamics.

6. DAREA set identification numbers should be unique with respect to all static load set identification numbers.

7. In a non-superelement analysis, LID and TID are not allowed to both be blank. In a superelements analysis, they may both be blank as long as static loads are prescribed in the upstream superelements.

Bulk Data Entry: **MAT1** [1] - Material Property Definition, Form 1.

Description: Defines the material properties for linear, temperature-independent, isotropic materials.

Format:

1	2	3	4	5	6	7	8	9	10
MAT1	MID	E	G	NU	RHO	A	TREF	GE	+x
+x	ST	SC	SS	MCSID					

Field	Contents
MID	Material identification number (Integer > 0).
E	Young's modulus (Real or blank).
G	Shear modulus (Real or blank).
NU	Poisson's ratio ($-1.0 <$ Real ≤ 0.5 or blank).
RHO	Mass density (Real)
A	Thermal expansion coefficient (Real).
TREF	Reference temperature for the calculation of: (1) thermal loads, or (2) a temperature dependent thermal expansion coefficient (Real or blank) (See Remarks 12 and 13).
GE	Structural element damping coefficient (Real) (See Remark 12).
ST, SC, SS	Stress limits for tension, compression, and shear (Real). (Used only to compute margins of safety in certain elements; they have no effect on the computational procedures.) See Sections 1.3.2 and 1.3.3.
MCSID	Material Coordinate System identification number (used only for CURV module processing) (Integer ≥ 0 or blank).

Remarks:

1 The material identification number must be unique for all MAT1, MAT2, MAT3 and MAT9 entries.

2 MAT1 materials may be made temperature dependent by the use of the MATT1 entry. In SOLution 66, linear and nonlinear elastic material properties in the residual structure will be updated as prescribed under the TEMPERATURE Case Control command.

3 The mass density, RHO, will be used to automatically compute mass for all structural elements.

4   Weight density may be used in field 6 if the value 1/g is entered on the PARAM entry WTMASS, where g is the acceleration of gravity (see Section 3.1.5).

5   MCSID must be nonzero if the CURV module is used to calculate stresses or strains at grid points on plate and shell elements only.

6   To obtain the damping coefficient, GE, multiply the critical damping ratio C/C_o by 2.0.

7   Either E or G must be specified (i.e., nonblank).

8   If any one of E, G, or NU is blank, it will be computed to satisfy the identity $E = 2(1+NU)G$; otherwise, values supplied by the user will be used. If any are temperature-dependent, then this calculation is only made for initial values of E, G, and NU. See remarks under MATT1 entry description.

9   If E and NU or G and NU are both blank, they will both be given the value 0.0.

10   Implausible data on one or more MAT1 entries will result in a warning message. Implausible data is defined as any $E < 0.0$ or $G < 0.0$ or $NU > 0.5$ or $NU < 0.0$ or $|1 - E/(2(1+NU)G)| > 0.01$ (except for cases covered by Remark 9).

11   It is strongly recommended that only two of the three values E, G, and NU be input. The three values may be input independently on the MAT2 entry. See the Handbook for Linear Static Analysis, Section 2.4.

12   TREF and GE are ignored if this entry is referenced by a PCOMP entry.

13   TREF is used for two different purposes:

a.   In SOLution 66, it is used only for the calculation of a temperature-dependent thermal expansion coefficient. (The reference temperature for the calculation of the thermal loads is obtained from the TEMPERATURE (INITIAL) set selection.)

b.   In all SOlutions except 66, it is used only as the reference temperature for the calculation of thermal loads. (TEMPERATURE(INITIAL) may be used for this purpose, but then TREF must be blank.)

Bulk Data Entry: **MOMENT** [7] - Static Moment.

Description: Defines a static moment at a grid point by specifying a vector.

Format:

1	2	3	4	5	6	7	8	9	10
MOMENT	SID	G	CID	M	N1	N2	N3		

Field	Contents
SID	Load set identification number (Integer > 0).
G	Grid point identification number (Integer > 0).
CID	Coordinate system identification number (Integer $\geq$ 0).
M	Scale factor (Real).
Ni	Components of vector measured in coordinate system defined by CID (Real; at least one nonzero component).

Remarks:

1   The static moment applied to grid point G is given by $m = M*(N1, N2, N3)$.

2   Load sets must be selected in the Case Control Section (LOAD = SID) to be selected by MSC/NASTRAN.

3   A CID of zero or blank references the basic coordinate system.

<u>Bulk Data Entry</u>: **MPC** [23] - Multipoint Constraint.

<u>Description</u>: Defines a multipoint constraint equation of the form:

$$\Sigma_i A_i u_i = 0$$

<u>Format</u>:

1	2	3	4	5	6	7	8	9	10
MPC	SID	G	C	A	G	C	A	blk	+x
+x	blk	G	C	A	etc.				

<u>Field</u>	<u>Contents</u>
SID	Set identification number (Integer > 0).
G	Identification number of grid or scalar point (Integer > 0).
C	Component number - any one of the digits 1-6 in the case of geometric grid points; blank or zero in the case of scalar points (Integer).
A	Coefficient (Real; the first A must be nonzero).

<u>Remarks</u>:

1   The first coordinate in the sequence is assumed to be the dependent coordinate. A dependent degree of freedom assigned by one MPC entry cannot be assigned dependent by another MPC entry or by a rigid element.

2   Forces of multipoint constraint are not recovered, except in SOL 24 with RFALTER RF24D24. See Section 3.5.

3   Multipoint constraint sets must be selected in the Case Control Section (MPC=SID) to be used by MSC/NASTRAN.

4   The m-set coordinates specified on this entry may not be specified on other entries that define mutually exclusive sets. See Section 1.4.1 for a list of these entries.

Bulk Data Entry: **PBAR** [7] - Simple Beam Property.

Description: Defines the properties of a simple beam (BAR) which is used to create bar elements via the CBAR entry.

Format:

1	2	3	4	5	6	7	8	9	10
PBAR	PID	MID	A	I1	I2	J	NSM	blk	+x
+x	C1	C2	D1	D2	E1	E2	F1	F2	+y
+y	K1	K2	I12						

Field	Contents
PID	Property identification number (Integer > 0).
MID	Material identification number (Integer > 0).
A	Area of bar cross-section (Real).
I1,I2, I12	Area moments of inertia (Real) ($I1 \geq 0.$, $I2 \geq 0.$, $I1*I2 > I12^2$).
J	Torsional constant (Real).
NSM	Nonstructural mass per unit length (Real).
K1, K2	Area factors for shear (Real).
Ci, Di, Ei, Fi	Stress recovery coefficients (Real).

Remarks:

1   For structural problems, PBAR entries may only reference MAT1 material entries.

2   See Section 1.3.2 for a discussion of bar element geometry.

3   For heat transfer problems, PBAR entries may only reference MAT4 or MAT5 material entries.

4   The transverse shear stiffness in planes 1 and 2 are (K1)AG and (K2)AG, respectively. The default values for K1 and K2 are infinite; in other words, the transverse shear flexibilities are set equal to zero. K1 and K2 are ignored if I12 is different than zero.

5   The stress recovery coefficients C1 and C2, etc., are the y and z coordinates in the BAR element coordinate system of a point at which stresses are computed. Stresses are computed at both ends of the BAR.

<u>Bulk Data Entry</u>: **PLOAD** - Static Pressure Load.

<u>Description</u>: Defines a static pressure load on a triangular or quadrilateral element.

<u>Format</u>:

1	2	3	4	5	6	7	8	9	10
PLOAD	SID	P	G1	G2	G3	G4			

<u>Field</u>	<u>Contents</u>
SID	Load set identification number (Integer > 0).
P	Pressure (Real).
Gi	Grid point identification numbers (Integer > 0; G4 may be zero).

<u>Remarks</u>:

1 The grid points define either a triangular or a quadrilateral surface to which a pressure is applied. If G4 is zero or blank, the surface is triangular.

2 In the case of a triangular surface, the assumed direction of the pressure is computed according to the right-hand rule using the sequence of grid points G1, G2, G3 as illustrated in the figure. The total load on the surface, AP, is divided into three equal parts and applied to the grid points as concentrated loads. A minus sign in field 3 reverses the direction of the load.

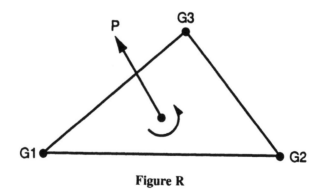

Figure R

3   In the case of a quadrilateral surface, the grid points G1, G2, G3 and G4 should form a consecutive sequence around the perimeter. The right-hand rule is applied to find the assumed direction of the pressure. Four concentrated loads are applied to the grid points in approximately the same manner as for a triangular face. The following specific procedures are adopted to accommodate irregular and/or warped surfaces:

a.   The surface is divided into two sets of overlapping triangular surfaces. Each triangular surface is bounded by two of the sides and one of the diagonals of the quadrilateral.

b.   One-half of the pressure is applied to each triangle which is then treated in the manner described in Remark 2.

4   Load sets must be selected in the case Control Section to be used by MSC/NASTRAN.

<u>Bulk Data Entry</u>: **PLOAD1** [7] - Applied Loads on BAR, BEAM or BEND elements.

<u>Description</u>: Defines concentrated, uniformly distributed, or linearly distributed applied loads to the BAR or BEAM elements at user chosen points along the axis. For the BEND element, only distributed loads over the entire length can be defined.

<u>Format</u>:

1	2	3	4	5	6	7	8	9	10
PLOAD1	SID	EID	TYPE	SCALE	X1	P1	X2	P2	

<u>Field</u>	<u>Contents</u>
SID	Load set identification number (Integer > 0).
EID	BAR, BEAM or BEND element identification number (Integer > 0).
TYPE	Load type, one of the following BCD values: FX, FY, FZ, FXE, FYE, FZE, MX, MY, MZ, MXE, MYE, MZE.
SCALE	Determines scale factor for X1, X2. Must be one of the following BCD values: LE, FR, LEPR, FRPR.
X1, X2	Distances along BAR, BEAM or BEND element axis from end A. (X2 may be blank or real, $X2 \geq X1 \geq 0$).
P1, P2	Load factors at positions X1, X2 (Real or blank).

<u>Remarks</u>:

1. If X1 is different than X2, a linearly varying distributed load will be applied to the element between positions X1 and X2, having an intensity per unit length of bar equal to P1 at X1 and equal to P2 at X2 except as noted in Remarks 7 and 8 below.

2. If X2 is blank or equal to X1, a concentrated load of value P1 will be applied at position X1.

3. If P1 = P2 and X2 is different than X1, a uniform distributed load of intensity per unit length equal to P1 will be applied between positions X1 and X2 except as noted in Remarks 7 and 8 below.

4. Load TYPE symbols are used as follows to define loads:

   FX, FY or FZ: Force in the x, y, or z direction of the basic coordinate system

MX, MY, or MZ: Moment in the x, y, or z direction of the basic coordinate system

FXE, FYE, FZE: Force in the x, y, or z direction of the element's coordinate system

MXE, MYE or MZE: Moment in the x, y, or z direction of the element's coordinate system.

5   If SCALE=LE (length), the Xi values are actual distances along the BAR axis, and (if X1 is different than X2) Pi are load intensities per unit length of the BAR.

6   If SCALE=FR (fractional), the Xi values are the ratios of the distance along the axis to the total length, and (if X2 is different than X1) Pi are load intensities per unit length of the BAR.

7   If SCALE=LEPR (length-projected), the Xi values are actual distance along the BAR axis and (if X2 is different than X1) the distributed load is input in terms of the projected length of the BAR.

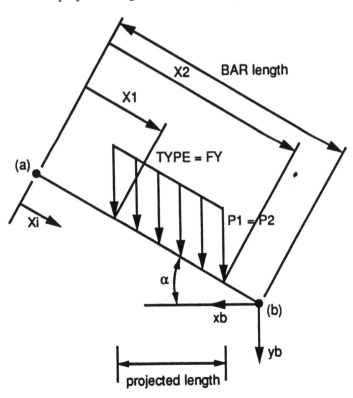

Figure S

If SCALE=LE, the total load applied to the bar is P1(X2-X1) in the yb direction.

If SCALE=LEPR, the total load applied to the bar is P1(X2-X1)cosα in the yb direction.

8   If SCALE=FRPR (fractional-projected), the Xi values are ratios of the actual distance to the length of the BAR, and if X1 is different than X2 the distributed load is input in terms of the projected length of the BAR.

9   Load sets must be selected in the Case Control Section (LOAD=SID) to be used by MSC/NASTRAN.

10   Element identification numbers for BAR, BEAM, and BEND must be unique.

11   For the BEND, the following coordinate equivalences have to be made for the element coordinates

$$R_{elem} = X_{elem}$$

$$\theta_{elem} = Y_{elem}$$

12   Only distributed loads applied over the entire length of the BEND can be applied.

13   Projected loads are not applicable to the BEND.

14   Loads on BEAM elements defined with PLOAD1 entries are applied along the line of the shear centers.

Bulk Data Entry: **PLOAD2** [10] - Pressure Load on a Two-Dimensional Structural Element.

Description: Defines a uniform static pressure load applied to two-dimensional elements. Only QUAD4, SHEAR, or TRIA3 elements may have a pressure load applied to them via this entry.

Format:

1	2	3	4	5	6	7	8	9	10
PLOAD2	SID	P	EID	EID	EID	EID	EID	EID	

Alternate Format:

1	2	3	4	5	6	7	8	9	10
PLOAD2	SID	P	EID1	THRU	EID2	blk	blk	blk	

Field	Contents
SID	Load set identification number (Integer > 0).
P	Pressure value (Real).
EID, EIDi	Element identification number (Integer > 0; EID1 $<$ EID2).

Remarks:

1 EID must be 0 or blank for omitted entries.

2 Load sets must be selected in the Case Control Section (LOAD = SID) to be used by MSC/NASTRAN.

3 At least one positive EID must be present on each PLOAD2 entry.

4 If the alternate format is used, all elements EID1 through EID2 must be two-dimensional.

5 The direction of the pressure is computed according to the right-hand rule using the grid point sequence specified on the element entry. Refer to the PLOAD entry.

6 All elements references must exist.

7 Continuations are not allowed.

<u>Bulk Data Entry</u>: **PLOAD3** - Pressure Load on a face of a HEX8 or a HEX20 Solid Element.

<u>Description</u>: Defines a uniform static pressure load applied to a surface of a HEX8 or a HEX20 solid element.

<u>Format</u>:

1	2	3	4	5	6	7	8	9	10
PLOAD3	SID	P	EID1	G11	G12	EID2	G21	G22	

<u>Field</u>	<u>Contents</u>
SID	Load set identification number (Integer > 0).
P	Pressure value (Real, force per unit area).
EID1, EID2	Element identification number (Integer > 0).
Gij	Grid point identification number of two grid points at diagonally opposite corners of the face on which the pressure acts (Integer > 0).

<u>Remarks</u>:

1 Load sets must be selected in the Case Control section (LOAD=SID) to be used by MSC/NASTRAN.

2 At least one EID must be present on the PLOAD3 entry.

3 Computations consider the pressure to act positive inward on the specified face of the element.

Bulk Data Entry: **PLOAD4** [13] - Pressure Loads on the Face of Structural Elements.

Description: Defines a load on a face of a HEXA, PENTA, TETRA, TRIA3, TRIA6, QUAD4, QUAD8 or QUADR element.

Format:

1	2	3	4	5	6	7	8	9		10
PLOAD4	SID	EID	P1	P2	P3	P4	G1	G3 or G4		+x
+x	CID	N1	N2	N3						

Alternate Format:

1	2	3	4	5	6	7	8	9	10
PLOAD4	SID	EID1	P1	P2	P3	P4	THRU	EID2	+y
+y	CID	N1	N2	N3					

Field	Contents
SID	Load set identification number (Integer > 0).
EID, EIDi	Element identification number (Integer > 0; EID1 < EID2).
Pi	Load per unit surface area (Pressure) at the corners of the face of the element (Real or blank). (P1 is the default for P2, P3 and P4).
G1	Identification number of a grid point connected to a corner of the face. Required data for solid elements only (Integer or blank).
G3	Identification number of a grid point connected to a corner diagonally opposite to G1 on the same face of a HEXA or PENTA element. Required data for quadrilateral faces of HEXA and PENTA elements only (Integer or blank). G3 must be omitted for a triangular surface on a PENTA element.
G4	Identification number of the TETRA grid point located at the corner not on the face being loaded. This is required data and is used for TETRA elements only.
CID	Coordinate system identification number (Integer ≥ 0).
Nj	Components of vector measured in coordinate system defined by CID (Real). Used to define the direction (but not the magnitude) of the load intensity.

Remarks:

1   The continuation is optional. If fields 2, 3, 4 and 5 of the continuation are blank, the load is assumed to be pressure acting normal to the face. If these fields are not blank, the load acts in the direction defined in these fields. Note that, if CID is a curvilinear coordinate system, the direction of loading may vary over the surface of the element. The load intensity is the load per unit of surface area, not the load per unit of area normal to the direction of loading.

2   For solid elements the direction of positive pressure (defaulted continuation) is inward. The load intensity P1 acts at grid point G1 and load intensities P2, P3 (and P4) act at the other corners in a sequence determined by applying the right hand rule to the outward normal.

3   For plate elements the direction of positive pressure (defaulted continuation) is the direction of positive normal, determined by applying the right hand rule to the sequence of connected grid points. The load intensities P1, P2, P3 (and P4) act respectively at corner points G1, G2, G3 (and G4). (See plate connection entries.)

4   If P2, P3 (and P4) are blank fields, the load intensity is uniform and equal to P1. P4 has no meaning for a triangular face and may be left blank in this case.

5   Equivalent grid point loads are computed by linear (or bilinear) interpolation of load intensity, followed by numerical integration using isoparametric shape functions. Note that a uniform load intensity will not necessarily result in equal equivalent grid point loads.

6   G1 and G3 are ignored for TRIA3, TRIA6, QUAD4, QUAD8 and QUADR elements.

7   The alternate form is available only for TRIA3, TRIA6, QUAD4, QUAD8 and QUADR elements. The continuation may be used in the alternate form.

8   For triangular faces of PENTA elements, G1 is an identification number of a corner grid point that is on the face being loaded and the G3 or G4 field is left blank. For faces of TETRA elements, G1 is an identification number of a corner grid point that is on the face being loaded and G4 is an identification number of the corner grid point that is not on the face being loaded. Since a TETRA has only

four corner points, this point, G4, will be unique and
different for each of the four faces of a TETRA element.

9   For the QUADR element, only pressure which acts normal
to the element is computed properly. (Surface tractions are
not resolved into moments normal to the element.)

<u>Bulk Data Entry</u>: **PROD** [1] - Rod Property.

<u>Description</u>: Defines the properties of a rod which is referenced by the CROD entry.

1	2	3	4	5	6	7	8	9	10
PROD	PID	MID	A	J	C	NSM			

<u>Field</u>	<u>Contents</u>
PID	Property identification number (Integer > 0).
MID	Material identification number (Integer > 0).
A	Area of rod (Real).
J	Torsional constant (Real).
C	Coefficient to determine torsional stress (Real) (Default = 0.0).
NSM	Nonstructural mass per unit length (Real).

<u>Remarks</u>:

1. PROD entries must all have unique property identification numbers.

2. For structural problems, PROD entries may only reference MAT1 material entries.

3. For heat transfer problems, PROD entries may only reference MAT4 or MAT5 entries.

4. The formula used to compute torsional stress is

$$\tau = CM_\theta / J$$

where M_θ is the torsional moment.

Bulk Data Entry: **PSHELL** [10] - Shell Element Property.

Description: Defines the membrane, bending, transverse shear, and coupling of thin shell elements.

Format:

1	2	3	4	5	6	7	8	9	10
PSHELL	PID	MID1	T	MID2	12I/T3	MID3	TS/T	NSM	+x
+x	Z1	Z2	MID4						

Field	Contents
PID	Property identification number (Integer > 0).
MID1	Material identification number for the membrane (Integer > -1 or blank).
T	Default value for membrane thickness.
MID2	Material identification number for bending (Integer $\geq$ -1 or blank).
12I/T3	Bending stiffness parameter (Real or blank, default = 1.0).
MID3	Material identification number for transverse shear (Integer > 0 or blank), must be blank unless MID2 > 0).
TS/T	Transverse shear thickness divided by the membrane thickness (Real or blank, default = .833333)
NSM	Nonstructural mass per unit area (Real).
Zi	Fiber distances for stress computation. The positive direction is determined by the right-hand rule and the order in which the grid points are listed on the connection entry. (Real or blank, see Remark 11 for defaults).
MID4	Material identification number for membrane-bending coupling (Integer > 0 or blank, must be blank unless MID1 > 0 and MID2 > 0, may not equal MID1 or MID2) (See Remarks 6 and 13).

Remarks:

1   All PSHELL property entries must have unique identification numbers.

2   The structural mass is computed from the density using the membrane thickness and membrane material properties.

3   The results of leaving an MID field blank (or MID2 = -1) are:

MID1   No membrane or coupling stiffness

MID2   No bending, coupling, or transverse shear stiffness

MID3   No transverse shear flexibility

MID4   No bending-membrane coupling

4   The continuation card is not required.

5   The structural damping (for dynamics rigid formats) uses the values defined for the MID1 material.

6   The MID4 field should be left blank if the material properties are symmetric with respect to the middle surface of the shell. If the element centerline is offset from the plane of the grid points, but the material properties are symmetric, the preferred method for modeling the offset is by use of the ZOFFS field on the connection entry. Although the MID4 field may be used for this purpose, it involves laborious calculations that produce physically unrealistic stiffness matrices ("negative terms on factor diagonal") if done incorrectly.

7   This entry is used in connection with the CTRIA3, CTRIA6, CQUAD4 and CQUAD8 entries.

8   For structural problems, PSHELL entries may reference a MAT1, MAT2 or MAT8 material property entries.

9   If the transverse shear material, MID3, references a MAT2 data entry, then G33 must be zero. If MID3 references a MAT8 data entry, then G1, Z and G2, Z must not be zero.

10   For heat transfer problems, PSHELL entries may reference MAT4 or MAT5 material property entries.

11   The default for Z1 is $-T/2$, and for Z2 is $+T/2$. T is the local plate thickness, defined either by T on this entry, or by membrane thickness at connected grid points, if they are input on connection entries.

12   For plane strain analysis, set MID2=-1. Only MAT1 type data is allowed for MID1.

13   For the QUADR element, the MID4 field should be left blank because its formulation does not include membrane-bending coupling.

14   If the MIDi fields are greater than or equal to 10^8, then parameter NOCOMPS is set to $+1$, indicating that

composite stress data recovery is desired. (MIDi fields greater than 10^8 are produced by PCOMP entries).

15   For material nonlinear analysis, MID1 must be the same as MID2, unless a plane strain (MID2=-1) formulation is desired.

<u>Bulk Data Entry</u>: **PSOLID** [13] - Properties of HEXA, PENTA and TETRA Solid Elements.

<u>Description</u>: Defines the properties of solid elements. Referenced by CHEXA, CPENTA and CTETRA entries.

<u>Format</u>:

1	2	3	4	5	6	7	8	9	10
PSOLID	PID	MID	CORDM	IN	STRESS	ISOP			

<u>Field</u>	<u>Contents</u>
PID	Property identification number (Integer > 0).
MID	Identification number of a MAT1, MAT4, MAT5, or MAT9 entry (Integer > 0).
CORDM	Identification number of material coordinate system (Integer, Default = -1).
IN	Integration network (Integer, BCD, or blank) (See Remark 4).
STRESS	Location selection for stress output (Integer, BCD, or blank) (See Remark 4).
ISOP	Integration scheme (Integer, BCD, or blank) (See Remark 4).

<u>Remarks</u>:

1   Psolid entries must have unique ID numbers.

2   Either isotropic (MAT1 or MAT4) or anisotropic (MAT5 or MAT9) materials be referenced.

3   See the CHEXA, CPENTA or CTETRA entry for the definition of the element coordinate system. The material coordinate system may be the basic system (0), any defined system (Integer > 0) or the element coordinate system (-1 or blank).

4   The following tables indicate the allowed options and combinations of options. If a combinations not found in the table is used, then the program will issue a warning message and assume default values for all options.

    The recommendation is to use the default (blank). The effect of using non-default values is as follows:

a.   If there are no midside nodes, IN = 2 produces reasonable output. However, IN = 3 produces an overly stiff element.

b.   If there are any midside nodes, the default value is in effect IN=3 (see the first part of this remark), which is the recommended value. If IN=2 is used, the element is under-integrated and has modes of deformation that lead to no strain energy.

c.   A standard isoparametric integration is requested by specifying IN=2 or 3 and ISOP=1.

d.   Stress output computed at Gauss points (STRESS=1) is available only for 8-node HEXA, 6-node PENTA and 4-node TETRA.

5.   The CORDM option is not effective in stress output labeled "NONLINEAR STRESSES", where only the element coordinate system is used.

HEXA	IN (8 node fault: BUBBLE) (20 node default: THREE)	STRESS (Default: GRID)	ISOP (Default: REDUCED)	Integration
8 node	blank or 0 or BUBBLE	GAUSS or 1	blank or REDUCED	2x2x2 Reduced Shear with Bubble Function
		blank or GRID		2x2x2 Reduced Shear Only
	TWO or 2		FULL or 1	2x2x2 Standard Isoparametric
9-20 node	blank or 3 or THREE		blank or REDUCED	3x3x3 Reduced Shear Only
			FULL or 1	3x3x3 Standard Isoparametric

Figure T

PENTA	IN (8 node default: BUBBLE) (20 node default: THREE)	STRESS (Default: GRID)	ISOP (Default: REDUCED)	Integration
⊠	blank or 0 or BUBBLE	GAUSS or 1	blank or REDUCED	3x2 Reduced Shear with BUBBLE Function
	TWO or 2	blank or GRID		3x2 Reduced Shear Only
7-15 node	blank or 3 or THREE			3x7 Reduced Shear Only
6 node			FULL or 1	3x7 Standard Isoparametric

Figure U

TETRA	In (4 node default: TWO) (10 node default: THREE)	STRESS (Default: GRID)	ISOP (Default: FULL)	Integration
4 node	blank or TWO or 2	GAUSS or 1	blank or FULL	1 Point Standard Isoparametric
5-10 node	blank or THREE or 3	blank or GRID		5 Point Standard Isoparametric

Figure V

Bulk Data Entry: **RBE2** [26] - Rigid Body Element, Form 2.

Description: Defines a rigid body whose independent degrees of freedom are specified at a single grid point and whose dependent degrees of freedom are specified at an arbitrary number of grid points.

Format:

1	2	3	4	5	6	7	8	9	10
RBE2	EID	GN	CM	GM1	GM2	GM3	GM4	GM5	+xx
+xx	GM6	GM7	GM8	etc.					

Field	Contents
EID	Identification number of rigid elements.
GN	The grid point to which all six <u>independent</u> degrees of freedom for the element are assigned (Integer > 0).
CM	Component number of the <u>dependent</u> degrees of freedom in the <u>global</u> coordinate system at grid points GM1, GM2, etc. The components are indicated by any of the digits 1-6 with no embedded blanks (Integer > 0).
GMi	Grid points at which dependent degrees of freedom are assigned.

Remarks:

1. The components indicated by CM are made dependent (members of the {um} set) at all grid points, GMi.

2. Dependent degrees of freedom assigned by one rigid element may not also be assigned dependent by another rigid element or by a multipoint constraint.

3. Element identification numbers must be unique.

4. Rigid elements, unlike MPC's, are <u>not</u> selected through the Case Control Section.

5. Forces of multipoint constraint are not recovered, except in SOL 24 with RFALTER RF24D24. See Section 3.5.

6. Rigid elements are ignored in heat transfer problems.

7. See Section 2.10 of the MSC/NASTRAN Application Manual for a discussion of rigid elements.

8. The m-set coordinates specified on this entry may not be specified on other entries that define mutually exclusive sets. See Section 1.4.1 for list of these entries.

<u>Bulk Data Entry</u>: **RLOAD1** [21] - Frequency Response Dynamic Load, Form 1.

Description: Defines a frequency dependent dynamic load of the form

$$\{P(f)\} = \{A[C(f) + i\ D(f)]\}\ exp(i\{\theta - 2\pi f\tau\})$$

for use in frequency response problems.

<u>Format</u>:

1	2	3	4	5	6	7	8	9	10
RLOAD1	SID	DAREA	DELAY	DPHASE	TC	TD			

<u>Field</u>	<u>Contents</u>
SID	Set identification number (Integer > 0).
DAREA	Identification number of DAREA entry which defines A (Integer > 0).
DELAY	Identification number of DELAY entry set which defines τ (Integer > 0).
DPHASE	Identification number of DPHASE entry set which defines θ (Integer > 0).
TC	Set identification number of TABLEDi entry which gives C(f) (Integer $\geq$ 0; TC + TD > 0).
TD	Set identification number of TABLEDi entry which gives D(f) (Integer $\geq$ 0; TC + TD > = 0).

<u>Remarks</u>:

1. If any of DELAY, DPHASE, TC or TD are blank or zero, the corresponding τ, θ, C(f), or D(f) will be zero.

2. Dynamic load sets must be selected in the Case Control Section (DLOAD = SID) to be used by MSC/NASTRAN.

3. RLOAD1 loads may be combined with RLOAD2 loads <u>only</u> by specification on a DLOAD entry. That is, the SID on a RLOAD1 entry may not be the same as that on a RLOAD2 entry.

4. SID must be unique for all RLOAD1, RLOAD2, TLOAD1 and TLOAD2 entries.

Bulk Data Entry: **SPC** [3] - Single-Point Constraint.

Description: Defines sets of single-point constraints and enforced displacements.

Format:

1	2	3	4	5	6	7	8	9	10
SPC	SID	G	C	D	G	C	D		

Field	Contents
SID	Identification number of single-point constraint set (Integer > 0).
G	Grid or scalar point identification number (Integer > 0).
C	Component number of global coordinate ($6 \geq$ Integer ≥ 0; up to six unique digits may be placed in the field with no embedded blanks.)
D	Value of enforced displacement for all coordinates designated by G and C (Real).

Remarks:

1. Coordinates specified on this entry form members of a mutually exclusive set. They may not be specified on other entries that define mutually exclusive sets. See Section 1.4.1 for a list of these entries.

2. Single-point forces of constraint are recovered during stress data recovery.

3. Single-point forces of constraint sets must be selected in the case Control Section (SPC = SID) to be used by MSC/NASTRAN.

4. From one to twelve single-point constraints may be defined on a single entry.

5. SPC degrees of freedom may be redundantly specified as permanent constraints on the GRID entry.

6. Continuations are not allowed.

<u>Bulk Data Entry</u>: **SPC1** [3] - Single-Point Constraint, Alternate Form.

<u>Description</u>: Defines sets of single-point constraints.

<u>Format</u>:

1	2	3	4	5	6	7	8	9	10
SPC1	SID	C	G1	G2	G3	G4	G5	G6	+x
+x	G7	G8	G9	etc.					

<u>Alternate Format</u>:

1	2	3	4	5	6	7	8	9	10
SPC1	SID	C	GID1	THRU	GID2				

<u>Field</u>	<u>Contents</u>
SID	Identification number of single-point constraint set (Integer > 0).
C	Component number of global coordinates (any unique combination of the digits 1-6 with no embedded blanks) when point identification numbers are grid points; must be null if point identification numbers are scalar points.
Gi, GIDi	Grid or scalar point identification number (Integer > 0).

<u>Remarks</u>:

1.   Note that enforced displacements are <u>not</u> available via this entry. As many continuations as desired may appear when "THRU" is not used.

2.   Coordinates specified on this entry form members of a mutually exclusive set. They may not be specified on other entries that define mutually exclusive sets. See Section 1.4.1 for a list of these entries.

3.   Single-point constraint sets must be selected in the Case Control Deck (SPC=SID) to be used by MSC/NASTRAN.

4.   SPC degrees of freedom may be redundantly specified as permanent constraints on the GRID entry.

5.   If the alternate form is used, points in the sequence GID1 through GID2 are not required to exist. Points which do not exist will collectively produce a warning message but will otherwise be ignored.

Bulk Data Entry: **TABDMP1** [20] - Modal Damping Table.

Description: Defines modal damping as a tabular function of frequency.

Format:

1	2	3	4	5	6	7	8	9	10
TABDMP1	ID	TYPE	blk	blk	blk	blk	blk	blk	+x
+x	f1	g1	f2	g2	f3	g3	f4	g4	

Field	Contents
ID	Table identification number (Integer > 0).
TYPE	BCD data word which indicates the type of damping units, "G", "CRIT", "Q", or blank. Default is "G".
fi	Frequency value in cycles per unit time (Real ≥ 0.0).
gi	Damping value.

Remarks:

1. The fi must be either ascending or descending order but not both.

2. Jumps between two points ($f_i = f_{i+1}$) are allowed, but not at the end points.

3. At least two entries must be present.

4. Any, fi, gi entry may be ignored by placing the BCD string "SKIP" in either of two fields used for that entry.

5. The end of the table is indicated by the existence of the BCD string "ENDT" in either of the two fields following the last entry. An error is detected if any continuations follow the entry containing the end-of-table flag "ENDT".

6. The TABDMP1 mnemonic infers the use of the algorithm

$$g = g_T(F)$$

where F is input to the table and g is returned. The table look-up $g_T(F)$ is performed using linear interpolation within the table and linear extrapolation outside the table using the last two end points at the appropriate table end. At jump points the average $g_T(F)$ is used. There are no error returns from this table look-up procedure.

7. Modal damping tables must be selected in the Case Control Section (SDAMP = ID) to be used by the program.

8   This form of damping is used only in modal formulations of complex eigenvalue analysis, frequency response analysis, or transient response analysis. The type of damping used ("structural", that is, displacement dependent or "viscous", that is velocity dependent) depends on the solution sequence. See Section 1.6.2 for the equations used.

9   PARAM, KDAMP may be used in Solution Sequences which perform modal frequency and modal complex analysis, to select the type of damping.

KDAMP = 1 (Default)   B Matrix

KDAMP = -1 $(1 + ig)K$ Matrix

See Section 3.3.13 for a full explanation.

10   If TYPE is "G" or blank, the damping values g1, g2, etc. are in units of equivalent viscous dampers,

$$b_i = (g_i/\omega_i)K_i$$

(See Section 3.3.12) If TYPE is "CRIT", the damping values g1, g2, etc. are in the units of fraction of critical damping, C/C_o. If TYPE is "Q", the damping values g1, g2, etc. are in the units of the amplification or quality factor, Q. These constants are related by the following equations:

$$C/C_o = g/2,$$

$$Q = 1/(C/C_o)$$

or

$$Q = 1/g$$

11   If this entry is stored in the data base, then repunched using SOL 60, the regenerated entries will all be converted to the equivalent "G"-type.

Bulk Data Entry: **TABLED1** [18] - Dynamic Load Tabular Function, Form 1.

Description: Defines a tabular function for use in generating frequency-dependent and time-dependent dynamic loads.

Format:

1	2	3	4	5	6	7	8	9	10
TABLED1	ID	blk	blk	blk	blk	blk	blk	blk	+x
+x	x1	y1	x2	y2	x3	y3	x4	y4	

Field	Contents
ID	Table identification number (Interger > 0).
xi, yi	Tabular entries (Real).

Remarks:

1 The xi must be in either ascending or descending order but not both.

2 Jumps between two points $(x_i = x_{i+1})$ are allowed, but not at the end points.

3 At least two entries must be present.

4 Any x-y entry may be ignored by placing the BCD string "SKIP" in either of the two fields used for that entry.

5 The end of the table is indicated by the existence of the BCD string "ENDT" in either of the two fields following the last entry. An error is detected if any continuations follow the entry containing the end-of-table flag "ENDT".

6 Each table mnemonic infers the use of a specific algorithm. For TABLED1 type tables, this algorithm is

$$Y = y_T(X)$$

where X is the input to the table and Y is returned. The table look-up $y_T(x)$, $x = X$ is performed using linear interpolation within the table and linear extrapolation outside the table using the last two end points at the appropriate table end. At jump points the average $y_T(x)$ is used. There are no error returns from this table look-up procedure.

7 Linear extrapolation is not used for Fourier Transform methods. The function is zero outside the range.

Bulk Data Entry: **TEMP** [4] - Grid Temperature Field.

Description: Defines temperature at grid points for determination of (1) Thermal Loading; (2) Temperature-dependent material properties; and (3) Stress recovery.

Format:

1	2	3	4	5	6	7	8	9	10
TEMP	SID	G	T	G	T	G	T	blk	

Field	Contents
SID	Temperature set identification number (Integer > 0).
G	Grid point identification number (Integer > 0).
T	Temperature (Real).

Remarks

1   Temperature sets must be selected in the Case Control Section (TEMP = SID) to be used by MSC/NASTRAN.

2   From one to three grid point temperature may be defined on a single entry.

3   If thermal effects are requested, all elements must have a temperature field defined either directly on a TEMPP1, TEMPP2, or TEMPRB entry or indirectly as the average of the connected grid point temperatures defined on the TEMP or TEMPD entries. Directly defined element temperatures always take precedence over the average of grid point temperatures.

4   If the element material is temperature dependent, its properties are evaluated at the average temperature.

5   Average element temperatures are obtained as a simple average of the connecting grid point temperatures when no element temperature data are defined.

6   Set ID must be unique with respect to all other LOAD type entries if TEMP (LOAD) is specified in the Case Control Deck.

Bulk Data Entry: **TIC** [19] - Transient Initial Condition.

Description: Defines values for the initial conditions of variables used in transient analysis. Both displacement and velocity values may be specified at independent degrees of freedom of the structural model.

Format:

1	2	3	4	5	6	7	8	9	10
TIC	SID	G	C	U0	V0				

Field	Contents
SID	Set identification number (Integer > 0).
G	Grid or scalar or extra point identification number (Integer > 0).
C	Component number (null or zero for scalar or extra points, any one of the digits 1-6 for a grid point).
U0	Initial displacement value (Real).
V0	Initial velocity value (Real).

Remarks:

1   Transient initial condition sets must be selected in the Case Control Section (IC=SID) to be used by MSC/NASTRAN.

2   If no TIC set is selected in the Case Control section, all initial conditions are assumed zero.

3   Initial conditions for coordinates not specified on TIC entries will be assumed zero.

4   Initial conditions may be used only in direct formulation. In a modal formulation the initial conditions are all zero.

5   For heat transfer, TEMP entries are used for initial conditions.

<u>Bulk Data Entry</u>: **TLOAD1** [18] - Transient Response Dynamic Load, Form 1.

<u>Description</u>: Defines a time dependent dynamic load or enforced motion of the form

$$\{P(t)\} = \{AF(t-\tau)\}$$

for use in a transient response problem.

<u>Format</u>:

1	2	3	4	5	6	7	8	9	10
TLOAD1	SID	DAREA	DELAY	TYPE	TID	blk	blk	blk	

Field	Contents
SID	Set identification number of DAREA entry set or a thermal load set (in heat transfer analysis) which defines A (Integer > 0).
DAREA	Identification number of DELAY entry set which defines τ (Integer $\geq$ 0).
TYPE	Defines the nature of the dynamic excitation (Integer 0, 1, 2, 3 or blank). (See Remark 2).
TID	Identification number of TABLEDi entry which gives $F(t-\tau)$ (Integer > 0).

<u>Remarks</u>:

1   If DELAY is blank or zero, τ will be zero.

2   The nature of the dynamic excitation is defined in field 5 in accordance with the following table.

Integer	Excitation Function
0 or blank	Force or Moment
1	Enforced Displacement
2	Enforced Velocity
3	Enforced Acceleration

See Section 3.5.4 of the MSC/NASTRAN Handbook for Dynamic Analysis regarding the use of the enforced motion options. Note that "large masses" must be used for enforced motion. For heat transfer problems, Field 5 must be blank.

3   Dynamic load sets must be selected in the Case Control Section (DLOAD = SID) to be used by MSC/NASTRAN.

4   TLOAD1 loads may be combined with TLOAD2 loads only by specification on a DLOAD entry. That is, the SID on a TLOAD1 entry may not be the same as that on a TLOAD2 entry.

5   SID must be unique for all TLOAD1, TLOAD2, RLOAD1 and RLOAD2 entries.

6   Field 3 may reference sets containing QHBDY, QBDY1, QBDY2, QVECT, and QVOL entries when using the heat transfer option.

7   If the heat transfer option is used, the referenced QVECT data entry may also contain references to functions of time, and therefore A may be a function of time.

8   Fourier analysis will be used if this is selected in an aeroelastic response problem.

Bulk Data Entry: **TLOAD2** [20] - Transient response Dynamic Load, Form 2.

Description: Defines a time-dependent dynamic load or enforced motion of the form

$$\{P(t)\} = \{0\}, \text{ if } t_* < 0 \text{ or } t_* > \text{T2-T1}$$

and

$$\{P(t)\} = At_*^B \exp(Ct) \cos(2\pi Ft_* + P), \text{ if } 0 \le t_* \le \text{T2-T1}$$

for use in a transient response problem where $t_* = t - T1 - \tau$.

Format:

1	2	3	4	5	6	7	8	9	10
TLOAD2	SID	DAREA	DELAY	TYPE	T1	T2	F	P	+xx
+xx	C	B							

Field	Contents
SID	Set identification number (Integer > 0).
DAREA	Identification number of DAREA entry set or a thermal load set (in heat transfer analysis) which defines A (Integer > 0).
TYPE	Defines the nature of the dynamic excitation (Integer 0, 1, 2, 3 or blank). (See Remark 2.)
T1	Time constant (Real $\ge$ 0.0).
T2	Time constant (Real, T2 > T1).
F	Frequency in cycles per unit time (Real $\ge$ 0.0).
P	Phase angle in degrees (Real).
C	Exponential coefficient (Real).
B	Growth coefficient (Real).

Remarks:

1. If DELAY is blank or zero, τ will be zero.

2. The nature of the dynamic excitation is defined in field 5 in accordance with the following table.

Integer	Excitation Function
0 or blank	Force or Moment
1	Enforced Displacement
2	Enforced Velocity
3	Enforced Acceleration

See Section 3.5.4 of the MSC/NASTRAN Handbook for Dynamic Analysis regarding the use of the enforced motion options. Note that "large masses" must be used for enforced motion. For heat transfer problems, Field 5 must be blank.

3   Dynamic load sets must be selected in the Case Control Section (DLOAD = SID) to be used by MSC/NASTRAN.

4   TLOAD1 loads may be combined with TLOAD2 loads only by specification on a DLOAD entry. That is, the SID on a TLOAD1 entry may not be the same as that on a TLOAD2 entry.

5   SID must be unique for all TLOAD1, TLOAD2, RLOAD1 and RLOAD2 entries.

6   Field 3 may reference sets containing QHBDY, QBDY1, QBDY2, QVECT, and QVOL entries when using the heat transfer option.

7   If the heat transfer option is used, the referenced QVECT data entry may also contain references to functions of time, and therefore A may be a function of time.

8   Fourier analysis will be used if this is selected in an aeroelastic response problem.

9   The continuation is optional.

Bulk Data Entry: **TSTEP** [18] - Transient Time Step.

Description: Defines time step intervals at which a solution will be generated and output in transient analysis.

Format:

1	2	3	4	5	6	7	8	9	10
TSTEP	SID	N(1)	DT(1)	NO(1)	blk	blk	blk	blk	+x
+x	blk	N(2)	DT(2)	NO(2)	blk	blk	blk	blk	

Example:

1	2	3	4	5	6	7	8	9	10
TSTEP	2	10	.001	5					+z
+z		9	.01	1					

Field	Contents
SID	Set identification number (Integer > 0).
N(i)	Number of time steps of value DT(i) (Integer ≥ 1).
DT(i)	Time increment (Real > 0.0)
NO(i)	Skip factor for output (every NO(i)th step will be saved for output) (Integer > 0).

Remarks:

1. TSTEP entries must be selected in the Case Control section (TSTEP=SID) in order to be used by MSC/NASTRAN.

2. Note that the entry permits changes in the size of the time step during the course of the solution. Thus, in the example shown, there are 10 time steps of value 0.001 followed by 9 time steps of value .01. Also, the user has requested that output be recorded for t=0.0, .005, .01. .02, .03, etc.

3. See Sections 11.3 and 11.4 of The NASTRAN Theoretical Manual for a discussion of considerations leading to the selection of time steps.

4. In Aeroelastic Response problems, this entry is required only when TLOAD is requested, i.e., when Fourier methods are selected.

5. The maximum and minimum displacement at each time step and the SIL numbers of these variables can be printed by altering PARAM//DIAGON//30 before the transient modules TRD1 or TRHT, and PARAM//DIAGOFF//30

after module. This is useful for runs that terminate due to overflow or excessive run times.

6   The number of output time steps must be less than 21,840. If more steps are selected than this on the TSTEP entry, the output set may be reduced by the use of the skip factor. The total number of time steps must be less than 65,523.

Appendix III

Selected Options for Output Request (Case Control Deck)

DISPLACEMENT	requests the displacement for a set of physical points
VELOCITY	requests the velocity for a set of physical points
ACCELERATION	requests the acceleration for a set of physical points
ELFORCE or FORCE	requests the forces for a set of structural elements
STRESS or STRESSES	requests the stresses for a set of structural elements
ESE	requests the strain energy for a set of elements
SPCFORCES	requests the reaction forces for a set of elements
OLOAD	requests the applied loads

Appendix IV

Selected Options for Contour Plots

Stress Contour Plots (for QUAD4s and TRIA3s)

CONTOUR xx yy zz ww

xx = type of stress

MAJPRIN	major principal stress
MINPRIN	minor principal stress
MAXSHEAR	maximum shear stress
XNORMAL	x component of normal stress
YNORMAL	y component of normal stress
ZNORMAL	z component of normal stress
XYSHEAR	xy component of shear stress
XZSHEAR	xz component of shear stress
YZSHEAR	yz component of shear stress

yy = number of stress contours

EVEN pp, where pp is the number of contour plots (default is 10)

LIST a, b, c,..., where a, b, c, ..., are values of stress contours

zz = location of stress (Z1 and Z2 specified on PSHELL card)

Z1	distance Z1 from neutral plane (default)
Z2	distance Z2 from neutral plane
MAX	maximum of stress at Z1 and Z2
MID	average stress at Z1 and Z2

ww =   coordinate system

COMMON   plot stress contours in basic coordinate
                system (default)
LOCAL   plot stress contours in element coordinate
                system

Displacement Contour Plots

CONTOUR xx yy

xx =   type of displacement

XDISP   (x component of displacement in global
                coordinate system)
YDISP   (y component of displacement in global
                coordinate system)
ZDISP   (z component of displacement in global
                coordinate system)
MAGNIT   (magnitude of deformation)

yy =   number of displacement contours

EVEN pp   where pp is the number of contour plots
                (default is 10)
LIST a, b, c,..., where a, b, c, ..., are values of displacement
                contours

Appendix V

Use of Static Loads in Dynamic Analysis

Case Control Deck

.

.

.

DLOAD = xx$^{\&}$
LOADSET = yy

.

.

.

Bulk Data Deck

.

.

.

TLOADi, xx, zz,...
LSEQ, yy, zz, ss
GRAV, ss,... or MOMENT, ss,...or PLOAD, ss,...

.

.

.

Example A.

Case Control Deck

DLOAD = 200
LOADSET = 77

$^{\&}$ This DLOAD card can also point to a DLOAD in the Bulk Data Deck, which in turn can point to a TLOADi card.

Bulk Data Deck

TLOAD2, 200, 47, blank, blank, 0.0, 10.E + 9, 1.0, 0.0
LSEQ, 77, 47, 249
GRAV, 249, blank, 2000., 1.0, 0., 0.

This results in a "time dependent" gravitational field equal to

$$g = g(t) = 2000 \cos(2\pi t)$$

acting in the positive x-direction.

Example B.

Case Control Deck

DLOAD = 55
LOADSET = 73

Bulk Data Deck

TLOAD2, 55, 20, blank, blank, 0.0, 10.E + 9, 1.0, 0.0
LSEQ, 73, 20, 7777
PLOAD2, 7777, 250., 22

This results in a time dependent pressure

$$p = p(t) = 250 \cos(2\pi t)$$

acting on element number 22.

Appendix VI

Summary of Topics Covered in Each Problem

PART 1 - STATICS

Problem 1

Static analysis of a 2-D truss
> Basic structure of input file
> Basic plot commands
> ROD elements
> Displacement output

Problem 2

Static analysis of 2-D truss
> Different output options
> EPSILON (numerical conditioning number)

Problem 3

Enforced displacement problem
> Use of SPC cards
> More about plots

Problem 4

Static analysis of truss subjected to two loading conditions
> SUBCASE structure
> Thermal loads
> Enforced deformations

Problem 5

Combination of loading conditions in static analysis
> Use of SUBCOM and SUBSEQ cards
> More output options (SET card)

Problem 6

Computation of safety margin coefficients
> Continuation cards
> Combination of static loads with LOAD card
> Yield stress

Problem 7

Static analysis of 2-D beam structure
    Gravity loads
    Distributed loads (PLOAD1 card)
    BAR elements
    BAR coordinate system (V-vector, Planes 1 and 2)

Problem 8

Static analysis of 2-D beam structure
    Shear stiffness coefficient
    Influence of shear stiffness in beam deformations

Problem 9

Static analysis of 3-D beam structure
    BAR coordinate system (use of G0)
    Use of pin flags

Problem 10

Square plate subjected to a uniform load
    QUAD4 elements
    Scalar springs

Problem 11

Static analysis of plate structure
    TRIA3 elements
    Use of replicator
    Contour plots
    PLOT SHRINK option

Problem 12

Plane stress problem
    QUAD8 element
    Membrane features of plate elements

Problem 13

Solid block subjected to a uniform pressure
    HEXA elements
    Symmetric boundary conditions

PART 2 - DYNAMICS

Problem 14

Normal modes of beam structure
>Inverse power method
>More about plots
>Output interpretation
>Concentrated masses

Problem 15

Normal modes of beam structure
>Modified Givens method
>Lumped mass matrix

Problem 16

Normal modes of beam structure
>Coupled mass matrix
>Options for normal modes scaling

Problem 17

Normal modes of a free structure
>Rigid body modes
>Singularity table for mass matrix

Problem 18

Transient response of beam to impulse load
>Direct method for transient response
>Numerical integration parameters
>Specification of dynamic loads
>X-Y plots
>Viscous dampers

Problem 19

Free vibrations of a beam under the action of gravity
>"Static" loads in dynamic analysis
>Non-homogeneous initial conditions

Problem 20

Transient response of beam to dynamic loads
>Modal method for transient analysis
>Combination of dynamic loads
>Modal damping
>More options to define dynamic loads

Problem 21

Frequency response analysis
    Definition of harmonic loads
    SORT2 output
    More about plots (magnitude and phase versus frequency)

PART 3 - A FEW MODELING TIPS

Problem 22

Static analysis of arch
    Use of cylindrical coordinates

Problem 23

Beam on inclined support
    Multipoint constraints

Problem 24

Beam on inclined support
    Auxiliary coordinate systems

Problem 25

Beam structure with excentric connection
    Use of offset vector

Problem 26

Beam structure with excentric connection
    Use of rigid bars

Problem 27

Structure subjected to base acceleration
    Tips for base acceleration problems

Problem 28

Structure subjected to base acceleration
    "Big Mass" approach

Subject Index

35871206R00266

Made in the USA
San Bernardino, CA
06 July 2016